Scientific Truth and Statistical Method

Books of cognate interest

Studies in the history of statistics and probability
E. S. PEARSON and M. G. KENDALL

*Games, gods and gambling: A history of probability
and statistical ideas*
F. N. DAVID

Statistical papers of George Udny Yule
selected by A. STUART and M. G. KENDALL

An introduction to the theory of statistics
G. U. YULE and M. G. KENDALL

The advanced theory of statistics (3 vol.)   M. G. KENDALL and A. STUART

A course in theoretical statistics   N. A. RAHMAN

A first course in statistics   F. N. DAVID

Biomathematics (2 vol.)   CEDRIC A. B. SMITH

Mathematical model building in economics and industry
(2 vol.: First and Second Series)   ed. M. G. KENDALL

Econometric techniques and problems   C. E. V. LESER

Style and vocabulary: Numerical studies   C. B. WILLIAMS

Man and the chemical elements   J. N. FRIEND

Scientific Truth and Statistical Method

Marcello Boldrini

formerly Professor of Statistics in the University of
Rome and President of the International Statistical
Institute. President of Ente Nazionale Idrocarburi,
1962–1967

translated from the Italian by
Ruth Kendall

Griffin 18 20 London

Charles Griffin & Company Limited
42 Drury Lane, London, WC2B 5RX

Copyright © 1972

This translation from the Italian has been made,
by permission, from the late Professor Marcello
Boldrini's copyright work, *Teoria della Statistica*
(Second Edition, 1965)

First English edition published in 1972

229 × 152 mm, xiv × 264 pages
4 line illustrations, 10 tables
ISBN 0 85264 197 4

SET AND PRINTED IN GREAT BRITAIN BY
CHORLEY & PICKERSGILL LIMITED LEEDS

Marcello Boldrini (1890-1969)

Professor Marcello Boldrini had a most distinguished career in both university and business. He was formerly Professor of Statistics in the University of Rome, Professor Honoris Causa of the University of Rio de Janeiro, an Honorary Fellow of the Royal Statistical Society, a Member of the Italian Academy of Sciences (Lincei), of the Pontificia Academia Scientiarum, etc.

He received many scientific honours, including the supreme honour of being President of the International Statistical Institute. He was also President of ENI (Ente Nazionale Idrocarburi) from 1962 to 1967. This book represents the results of a lifetime's reflection on the theoretical and practical applications of statistics and scientific inference.

This translation was approved by Professor Boldrini except for part of the last chapter, which has been approved by one of his distinguished pupils, Professor A. Maros dell'Oro.

A*

Author's Preface

What can one say by way of introduction to this book? The answer is to be found in a trial which took place 2,000 years ago, when some immortal words were spoken. "Quid est veritas?" the perplexed Pontius Pilate asked himself after an interrogation which had left him full of anguish. Jesus, the accused, had already given an answer when he stated with authority, "Ego sum Veritas". In that tragic moment the answer reached but few hearts, but it set out on its way down the centuries.

This book is entirely concerned with the development of Pilate's question and ends by accepting, on the very last page, the answer of Jesus. Indeed, the opinion was once firmly held that scientific truth was something essential and predetermined, a hidden principle of the physical world and a difficult objective for studious minds to achieve stage by stage, through conjecture and experiment. That opinion has now been shown to be mistaken. By scientific truth is still meant, of course, a certainty, but a subjective one, transitory, a special relation between man and the world, adapting itself to the progress of knowledge and to changes of interpretation and of human requirements.

For more than a century, scientific criticism has, in fact, been reviewing many earlier opinions, and by now very little remains of the general categories in which it was once thought that all knowledge could be contained. The self-evidence of mathematical and physical axioms, the concepts of absolute space and time and of universal determinism have been rejected, together with the reality of physical causes and of mathematical laws of the universe. Man has neither wished nor been able to find substitutes for them, except in the quest of strict consistency on the abstract level, and in the idea of probability on the practical level. It is now accepted that concepts and theories must be modified — either through discussion or experiment — whenever they prove to be inadequate.

Modern epistemology recognizes that science stands on two cornerstones. One is classification, which eliminates or interprets mathematically the variability of events, the other is the common noun, which enables classified events to be expressed and communicated.

Indeed, both are parts of the structure of induction, since they consider events regardless of the time of occurrence, that is, they actualize the future and furnish a rational basis for scientific forecasting. The realization that this is so seems all the more important in that it denies the objective regularity of phenomena which many authors still more or less openly accept, despite the earlier, penetrating warnings of Mach. Regularity is confirmed, it is true, by innumerable past instances, but this in no wise logically justifies its continuance in the future. Instead, regularity and repetition become, by means of classification, subjective realities independent of time (or, if you prefer, located in a conventional, static time), until the necessity arises to modify the concepts and so to vary the linguistic expressions which describe them.

The verbal reality of the common noun, as well as of the propositions in which something is predicated of it, and of the arguments and scientific structures of which, in a certain sense, it becomes the protagonist, give to modern science the appearance of a discourse: and it is therefore said that scientific reality *is* language. A science thus constructed, which no longer starts from absolute principles but formulates its premises conventionally from time to time and develops them linguistically, with the sole obligation of not contradicting itself, has been likened aptly enough to the work of the mosaic-setter who builds up pictures of his own world from the jumble of tiles which he, or others for him, have made from natural materials.

In science — as opposed to art — it is methodic rigour, rather than imagination, which predominates. In the widest sense, all the sciences rely for rigour on the discipline of method, that is, on Statistics.

With Statistics, logical argument develops theoretical structures and the mathematics of the physical world; that is, Statistics is the basic instrument for describing and elaborating classified events. This is the origin of the universal function of Statistics (in formulating hypotheses, designing and analysing experiments) and also of its unique quality of being a formal element running throughout the scientific process. However, it is not an extraneous element taken over from some region outside the sciences themselves, but keeps pace with them and adapts its operational techniques accordingly, so that anyone who says "science" also says "Statistics", and vice versa — knowing that Statistics can no more be avoided than can logic, by which man thinks, or language, by which he expresses himself.

Such a concept may seem deceptive, but in fact it is not so. One need only consider the splendid achievements of present-day knowledge, on a scale previously unimaginable — from the abstract to the concrete, from the field of physics to that of biology and sociology — to realize that man, with research and reasoning, has enlarged and transformed the boundaries of the universe. If current science was misleading, we should, after so much destructive criticism of it, have reverted to the obscurity of remote times. On the contrary, the liberty created by modern knowledge has lent wings to talent, broken down cultural barriers, and offered to all — even to modest research workers — the opportunity of taking part in the enrichment of knowledge.

Naturally, in the relative truths of a science historically conditioned and essentially changeable, in which the possible predominates over the necessary, the probable over the determined, the transitory over the definitive and, in a certain sense also, creativity over objectivity, it would be vain to regret the disappearance of those universal positive foundations which past thinkers had so laboriously built up on the basis of logic, mathematics, and experiment. But it is precisely in this lack of certainty that the transcendent Truth re-emerges as free choice and refers us back to the words "Ego sum Veritas". I propose as a solution a metaphysical interpretation of the universe.

This is the meaning and conclusion of a book which is certainly too large an undertaking not to contain serious defects. But lengthy study and a great affection for my work have, perhaps, enabled me to keep within the bounds of modern scientific subjectivism.

ROME,
March, 1968

Note

The numbering of paragraphs in the text follows the decimal system. The first figure of each section refers to the chapter, the next to a subject (topic), and the following numbers — as far as possible — refer to its developments. Thus, 58 indicates, in the fifth chapter, the subject section 8, while 581 indicates a first development of the question treated in section 58.

In the following slightly abridged contents list the paragraph numbers are omitted. The page numbers indicate the beginning of a new subject section in each chapter (e.g. 5.1, 5.2 and so on).

Contents

3   Deduction and Induction   54

4 Probability and Statistics 121

1

Science and language

11 Appearance and interpretation of the physical world

"I have settled down to the task of writing these lectures and have drawn up my chair to my two tables. Two tables! . . . One of them has been familiar to me from earliest years. It is a commonplace object of that environment which I call the world. . . . It is a thing. . . . Table No. 2 is my scientific table. It is a more recent acquaintance and I do not feel so familiar with it. It does not belong to the world previously mentioned — that world which spontaneously appears around me when I open my eyes. . . . My scientific table is mostly emptiness. Sparsely scattered in that emptiness are numerous electric charges rushing about with great speed; but their combined bulk amounts to less than a billionth of the bulk of the table itself. Notwithstanding its strange constitution, it turns out to be an entirely efficient table. It supports my writing papers as satisfactorily as table No. 1; for when I lay the paper on it the little electric particles with their headlong speed keep on hitting the underside, so that the paper is maintained in shuttlecock fashion at a nearly steady level; but for me it always remains the table I write at every day."

Eddington wrote these words in 1927. And yet the scientist of our day would take a very different view; the table as it appears would be a thing but the scientific table would be quite a different entity, to which the name of "thing" no longer seems even pertinent: it is an idea and its linguistic formulation.

111 From classical to modern physics

Such a radical change in the way of representing what we call the external world — comparable only to that determined in the seventeenth century by the Galilean theory of motion, which upset the Aristotelian beliefs that had survived for 1,500 years[1] — is due to the recent introduction of concepts which have revolutionized physics and, as a result, have greatly transformed the entire epistemology of science.

---

[1] Cf. H. Butterfield, *The Origins of Modern Science, 1300–1800*, George Bell & Sons, London, 1949, Chapter 4.

1

Lord Kelvin (1824–1907) is quoted as saying in 1884, "I never satisfy myself until I can make a mechanical model of a thing. If I can make a mechanical model of it I can understand it. As long as I cannot make a mechanical model all the way through, I cannot understand it. . . . But I want to understand light as well as I can, without introducing things that we understand even less of."[1] Now the mechanical models to which Lord Kelvin referred, and which are derived directly from the *Principia* of Newton (1687), lost their earlier prestige of absolute certainty at the beginning of our century, because of thorough inquiry into two phenomena which Kelvin judged to be two dark clouds on classical physics, namely the Michelson–Morley experiment, which started Einstein on the path of relativity (1905), and the experiments on black-body radiation which led to Planck's quantum theory (1900).

112   The intellectual revolution of the present century

Bridgman emphasizes the profound gulf between the ancient and the new concepts of the world brought about by the theory of relativity and by quantum physics, maintaining that a similar revolution can never recur. Indeed — according to Bridgman — Einstein and Planck would have taught, contrary to what was believed in the past, that science is not founded on absolute and immutable principles, but that its axioms are only provisional working hypotheses, to be discarded without hesitation or regret and therefore without concern, as soon as they are at variance with the facts. Apropos of this, Giusti's lines come to mind: scoffing at the prediction that Italian poetry would become barbaric through giving up courtly language, he ends by saying, "And after all this, what happened? Manzoni came on the scene." Today, both physics and science in general can be revolutionized without fundamental disturbance.

Newton wrote: "True, Absolute and Mathematical Time, of itself and from its nature, flows uniformly without reference to anything external, and by another name is called duration." Taking away the emphasis of the capital letters and bringing time down from its secular throne, thereby reducing "absolute time" to the practical idea of a time-interval (section 233), scientific epistemology has adapted itself to various current theories, favouring the progress of those achievements

---

[1] Quoted by P. W. Bridgman, *The Logic of Modern Physics*, 1927.

which most genuinely express the present phase of intellectual development (section 221)[1].

12 What science is

It is difficult to explain briefly what this science is, whose foundations have turned into a sort of quicksands but which nevertheless erects its imposing structures. Quite inadequately, and in order to keep our bearings, we will say here that science is always a mental construct which starts from abstract postulates or is derived from experimental data. Therefore, even when it has a factual content, science cannot be reduced to a photographic description of the world, but is an interpretation and exposition of its manifestations, conformably with one or more theories.

These words contain obvious limitations. In the first place, they affirm that everything which is not a mental construct is foreign to science, whether abstract or based on fact. In particular, metaphysics remains outside science because it tries to solve, by quite different means, the ultimate problems regarding the soul and man's destiny. Note that this is not a condemnation of metaphysics made in the name of science (although that is the erroneous conclusion often accepted by modern epistemologists) but merely a delimitation — that is, a recognition that metaphysical problems are not scientific ones and should therefore be treated separately and with suitable methodology, which is not that of science[2].

The other limitation is that science always contains an element of the conceptual, sometimes purely abstract and sometimes more or less substantiated by facts. From this fundamental element of interpretation stems the precarious nature of the scientific construct, quite apart from possible errors or other shortcomings of the data used.

---

[1] L. Geymonat, *Filosofia e filosofia della scienza*, Feltrinelli, Milan, 1962. This work, which will often be quoted, is well informed on the concept of the historical truth of scientific thought and of scientific progress as a continual achievement of methodological rigour. For more general references see *Proceedings of the 1960 International Congress*, edited by E. Nagel, P. Suppès and A. Tarski, entitled: *Logic, Methodology and Philosophy of Science*, Stanford University Press, Stanford, 1962. The volume contains 63 essays by leading international writers, and many bibliographical references.

[2] M. Boldrini, "Causa vel Casus", *Statistica*, October–December, 1960. For a positive appreciation of metaphysics, see H. Scholz. *Abriss der Geschichte der Logik* (1931). Italian transl.: *Storia della logica*, Silva, Milan, 1962, p. 126.

Finally, just because ideas and facts are presented to other people in an intelligible way, suitable for discussion, there is added a further subjective element — the linguistic factor, which is the main instrument of communication.

121   Abstract and concrete sciences

From what has been said in the preceding section, it follows that there are abstract and concrete sciences, the former purely formal and the latter substantiated by fact. In both cases they are objective and are freed from arbitrariness by following rules chosen *a priori*.

As to the factual disciplines, one should bear in mind that their development and progress is closely linked to their field of application, to the instrumentation at their disposal, and to the level of sophistication which they have reached. All the factual sciences are subject to this process, from descriptive exposition to quantitative expression. Compare, for example, such disciplines as are still predominantly qualitative, such as geology, geography and linguistics, with others of a measurable and mathematical character, such as physics, chemistry, genetics and demography.

The abstract sciences, on the other hand, are constituted by mental systems with self-determined values, but they provide special instruments of investigation for the practical sciences. Mathematics and logic are good examples of formal disciplines which are important for research in the concrete sciences. I shall refer to this topic again later.

13   Linguistic communicability of science

To say that science is always communicated in ordinary language may seem, at first sight, to be a commonplace, but exactly the contrary is true. All subjects have their own nomenclature, in ignorance of which it is impossible to learn them. The example of doctors' language — obscure yet necessary, even if a trifle overdone — is known to all.

Apart from this, other and more serious questions arise from the need to communicate, without misunderstanding, both the thoughts of those who cultivate the abstract sciences and the experimental results obtained by natural scientists and technicians.

131   Misunderstandings deriving from the homonymy of language

Language is plainly homonymous: this comes about because there exist many more things than there are words, so that even the most ordinary words (such as impression, conduct, post, place) have numerous meanings. Hegel said that homonymy delighted him, and he justified this to himself by instancing the German word *aufheben*, which can take on the opposite meanings of "to preserve" and "to destroy". For this reason he saw in it a confirmation of the dialectic which is at the centre of his philosophy[1]. One could, however, object that in Hegelianism and the "romantic" philosophies it would be difficult to set up science on a rigorous basis (see also section 1431)[2]. Not even with the usual syntax can one always avoid ambiguity; and unless great care is taken, language lends itself ill to expressing scientific thoughts which must, above all, be consistent and free from contradiction[3].

132 Special nature of scientific language

The better to appreciate the importance of scientific language, one may consider Galileo's bitter controversy with the opponents of the "New Sciences" and of the new system of the world. He was clearly aware that the dispute about the perfection of the heavens was not a question of fact but of words, to the extent that what appeared to the naked eye as evidence was not comparable with what was seen through a telescope. He also maintained, for example, that the biblical narrative of the miracle performed by Joshua was a version adapted to the popular intelligence, and that a different version would be required for the rigorous description one would give of it when considering a heliocentric cosmogony. Even today, when the Copernican theory is never seriously in question, we are accustomed to say that the sun rises and

---

[1] Compare the preface by J. Ohana to the French edition of A. J. Ayer's *Language, Truth and Logic*, Flammarion, Paris, 1956.

[2] See M. Boldrini, "Bertrand Russell", in *La Scuola in Azione*, 1959–60, No. 9.

[3] The following instances are significant:

St. Luke (19, 3–4) relates that the publican Zacchaeus sought to see Jesus, "and could not for the press, because he was little of stature". Although the subject of the sentence is Zacchaeus, some ancient and modern writers have pretended to read into the text a reference to the stature of Our Lord.*

Many art critics almost wilfully cause misunderstanding in their expert comment. Here are two examples, concerned with famous people. About an obvious copy of a well-known painting in the Louvre it was said, "It is undoubtedly a portrait of (di) E. Vigée Lebrun", which, in Italian, can mean either a painting by, or a portrait of, her. And thus runs the comment on a work containing mannerisms of Raphael's: "It is certainly a painting done 'nello studio' of Raphael", which can mean either inspired by his works or painted in his studio, at least by a pupil of his.

* *Translator's note* — In the *New English Bible* this ambiguity is avoided.

sets, instead of using the correct scientific form, that the rotation of the earth makes the sun appear and disappear on the horizon[1].

133   Science versus linguistic rigour

The example of Galileo makes it clear that the development of science and the constitution of its language are correlated elements; a science whose statements and arguments are not strictly formulated is never communicable without ambiguity.

Geymonat, following Michel, states that the first mathematical researches of the Greeks were expressed in ordinary language, and that a technical language and a notation were gradually worked out as necessity arose[2].

From a study by A. Pasquinelli we learn that the change in the concepts introduced into chemistry by Lavoisier came about through a critical analysis of the language used by the chemists of his time. Above all, writes Pasquinelli, the terms "phlogiston" and "element" seemed to Lavoisier to be loaded with doubtful and obscure empirical references, liable to hinder inquiry until they could be exactly defined by experiment. The man who is often justly called the father of modern chemistry directed his work bearing in mind the idea of Condillac, that "une science n'est qu'une langue bien faite"[3].

14   Russell's philosophy of logical analysis

But if the common tongue remains indifferent to exact expression, linguistic accuracy, first used in the so-called exact sciences, has spread to all the others.

The first thinker to put language at the centre of a philosophic system was Bertrand (Lord) Russell (1872–1970). With him must rank the Austrian, Ludwig Wittgenstein (1889–1951) who, in a work which has become famous, systematically developed the theory of a scientific language.

---

[1] In the same chapter of the Book of Joshua as contains the miracle of the sun, there is another passage which might have given rise to an analogous philological dispute, if the author had not forestalled any misunderstanding. It runs thus: "The Lord cast down great stones from Heaven upon them. . . they were more which died from hailstones than they whom the children of Israel slew with the sword." Here the repetition makes it clear that the stones were in fact large hailstones.

[2] L. Geymonat, *Filosofia e filosofia della scienza*, p. 172; L. Geymonat, "Storia del pensiero scientifico contemporaneo", in *La Scuola in Azione*, 1960–61, No. 1.

[3] A. Pasquinelli, "Problemi attuali della scienza come problemi di linguaggio", in *La Scuola in Azione*, 1960–61, No. 4.

Russell refers to his own field of study as the philosophy of logical analysis. He mentions — though considering it exaggerated — a statement of Rudolf Carnap (1891–1970, third in importance in this subject), according to whom, provided that errors of syntax are avoided, any philosophical problem can be resolved or else shown to be inconsistent in itself[1].

According to Russell, every "description" is a statement which can be replaced by a proper or common noun. For example: "The Maid of Orleans who died on the pyre"; "the philosopher of logical analysis"; "the domestic feline"; "the amphibian which is proof against fire". These mean, respectively, Joan of Arc, Bertrand Russell, the cat, the salamander.

The descriptions of Saint Joan and the salamander, to take the first and last, seem to be of the same type, but in fact are not so. To the objection that an amphibian with such a faculty does not exist, someone who accepted the opinion of Pliny or believed the testimony of Benvenuto Cellini could reply: "what does not exist?" and the other assert: "the salamander which withstands fire", thus falling into the ambiguity of affirming a non-existing thing.

In science this cannot be tolerated. Indeed, the above assertion lends itself to double interpretation, that is: "I affirm this to be the salamander resistant to fire", or "it is the salamander resistant to fire of which I deny the existence". It is a syntactic, not a verbal homonymy, which science must reject in its own interests, and with the same strictness with which it expels logical and factual errors.

Russell maintains that, to avoid misunderstanding, the sentence ought to be expressed in the following way: "There does not exist an animal s (the salamander) such that, putting s in the place of x in the expression 'x is resistant to fire', renders it true." Thus, says Russell, must the scientist express himself, even supposing ordinary speech could tolerate, for practical purposes, such wording as this.

After that, one can imagine how the Bible would have told the story of Joshua if the Holy Scriptures had had a scientific character rather than historical and edifying.

On another level, Russell affirms that descriptions of the type "The Maid of Domrémy who died on the pyre" are the instruments by which one legitimately affirms existence. It is equivalent to the explicit expression: "The Maid of Domrémy exists: she died on the pyre." On

---

[1] Among the many works of the British philosopher should be mentioned at this point B. Russell, *History of Western Philosophy*, 2nd edn, Allen and Unwin, London, 1963, Chapter 21.

the other hand, whoever says, avoiding description: "Joan of Arc exists", would be guilty of bad grammar, or rather of bad syntax. Russell adds: "This clears up two millennia of muddled-headedness about 'existence', beginning with Plato's *Theaetetus*" (see also sections 35 *et seq.* and 4621)[1].

141   Wittgenstein and the identity of science and language

After World War I a bold step forward was taken by Wittgenstein with the affirmation of the identity of science and language. His *Tractatus logico-philosophicus*, published in 1921[2], is an attempt to expound logic axiomatically and has had an enormous influence, both on the subsequent course of research and on Russell himself. At Cambridge the latter was first Wittgenstein's maestro and then his friend[3].

A short collection of Wittgenstein's axioms will give a better account of his thought than a long description.

1	The world is all that is the case.
1.1	The world is the totality of facts, not of things.
2	What is the case, a fact, is the existence of states of affairs.
2.04	The totality of existing states of affairs is the world.
2.1	We picture facts to ourselves.
2.11	A picture presents a situation in logical space, the existence and non-existence of states of affairs.
2.2	A picture has logico-pictorial form in common with what it depicts.
3	A logical picture of fact is a thought.
3.01	The totality of true thoughts is a picture of the world.
3.11	We use the perceptible sign of a proposition (spoken or written, etc.) as a projection of a possible situation. The method of projection is to think out the sense of the proposition.

---

[1] B. Russell, *Introduction to Mathematical Philosophy*. 3rd edn, Allen and Unwin, London, 1924. See also H. Scholz, *Storia della logica*, p. 121.

[2] L. Wittgenstein, *Tractatus logico-philosophicus*, original text German, revised English edition by D. F. Pears and B. F. McGuiness, Routledge and Kegan Paul, 1961. Towards the end of his life Wittgenstein modified some of his views, not always with Russell's agreement. See in particular his *Philosophical Investigations*, Basil Blackwell, Oxford, 1963, and *The Blue and Brown Books*, 2nd edn, Basil Blackwell, Oxford, 1964.

[3] "I myself was very much influenced by his earlier doctrines. ... He was immensely impressive as he had fire and penetration and intellectual purity to a quite extraordinary degree." B. Russell, *Portraits from Memory*, London, Allen & Unwin, 1956, p. 29. Russell had expressed more or less the same concepts in his obituary of Wittgenstein.

3.3   Only propositions have sense; only in the nexus of a proposition does a noun have meaning.

4   A thought is a proposition with a sense.

4.001   The totality of propositions is language.

4.0031   All philosophy is a "critique" of language.

4.01   A proposition is a picture of reality.

4.21   The simplest kind of proposition, an elementary proposition, asserts the existence of a state of affairs.

These extracts, of course, give an inadequate idea of this great and difficult work[1], but they illustrate the point that, according to Wittgenstein's logic, the world becomes comprehensible through its linguistic representation, and that the significant propositions are thought — that is to say, pictures of reality in the logical sense[2].

---

[1] Entire books on the interpretation of this work have been published in recent years, such as the following: E. Stenius, *Wittgenstein's "Tractatus"*, *A Critical Exposition of the Main Lines of Thought*, Blackwell, Oxford, 1960. The author, in his long and patient interpretation, warns that in the *Tractatus* there are "propositions which I think I understand, and which I think are illuminating, stimulating and important". There follow, he adds, "propositions which I think I understand and almost with certainty I think are false and equivocal". Finally, there are others "which I do not understand and the value of which I am therefore unable to judge".

     J. Jörgensen ("Origini e sviluppi dell'empirismo logico", an article in *Neopositivismo e unità della scienza*, Bompiani, Milan, 1958, pp. 148–64) for his part states that the *Tractatus* "has in it something irritating because of its strange mixture of lucid clarity and obscure profundity". This book was discussed "from top to bottom" in the Vienna Circle (cf. section 143), and he regrets that "the comment on it by Friedrich Waismann, announced several years ago, was not published". This comment would have had great importance, because Waismann, together with Schlick, had discussed it at length with the author. Points of clarification are to be found in the anthology containing two biographies by Norman Malcolm and Georg Henrik von Wright, *Ludwig Wittgenstein* (1958), Italian edition, Bompiani, Milan, 1960.

[2] Although not accepted without qualification, this concept is nowadays the most widespread.

     "Whether the actual process of thinking or reasoning requires language or not is an open question. It may be that thinking requires the use of symbols of some sort, words or images or whatnot. We all feel a certain sympathy with the girl who was told to think before she spoke and replied: 'But how can I know what I think until I hear what I say?'. Perhaps all thinking does require words or some other kind of symbols. . . . We know, however, that the *communication* of any proposition or any argument requires symbols, and can be accomplished only by the use of language." Cf. I. M. Copi, *Introduction to Logic*, Macmillan, New York, 1957, p. 12.

142   Linguistic formalization in the sciences

The most evident and complete realization of what Wittgenstein affirms is to be found in the abstract sciences, in particular in mathematics and logic. In these subjects formalization is achieved with a truly autonomous language, having its own rules of syntax and a rigorous consequentiality. But one also finds certain practical subjects, such as chemistry, in which the formulae can represent rigorously and, so to speak, in a visible form, the hypothetical structure of the most complex molecular compounds and allow reasoning about such symbolism by precise rules of transformation. Genetics aims to reach a similar position and is gradually formulating a symbolic language for some of its problems.

The science–language identity appears to be less evident in those subjects which are expressed in ordinary language — but this is no more than appearance. By considering the functioning of instruments of research and measurement we realize that words have lost their former content, and that all previous expressions and relationships must be reconsidered. The introduction of mechanical clocks, and later of pendulum clocks, completely removed from our language the Roman division of daily time into hours, while with the coming of the thermometer, the division of tactile sensations into hot and cold has given way to measures based on the height of a column of mercury in a vacuum. It is evident that between the two latter concepts and hence between their ways of expressing temperature there is no connection at all.

1421   Evolution and ambiguity of scientific languages

Bridgman has subjected the word "length" to a profound analysis. He pointed out the different meanings the word takes on in order to express the simple ratio of a segment to a predetermined standard metre, or stellar distances calculated from the velocity of light, or the distance between the planes of a crystal, which are determined by reference to the wave theory of light[1]: and hence the absurdity of making comparisons between these heterogeneous measures when reduced to metres, in view of the different meanings which even the word "metre" can acquire in them.

Dingle has convincingly demonstrated that the wave–corpuscle paradox, so debated in contemporary physics, is purely a linguistic

---

[1] P. H. Bridgman, *loc. cit.*, pp. 9 *et seq.*

question[1]. The electron is a conceptual entity, which from one aspect is regarded as a wave and from another as a corpuscle. The use of the same name is purely historical and does not imply an objective transmutation. If, historically, both light-waves and sound-waves had been termed "light", we could have deluded ourselves that, at times, such waves become sensible to the eye and at other times to the ear.

Reichenbach's conclusions some years later (1951) were similar to those of Dingle. Reichenbach[2], considering the anomalous behaviour of subatomic entities, wrote that "the controversy between the adherents of the wave and the corpuscle theories has been transformed into a duality of interpretations.

"Whether the constituents of matter are waves or particles is a question concerning unobservables; and unobservables of atomic dimensions, unlike those of the world at large, cannot be uniquely determined by the postulates of a normal system — because no such system exists." Shortly afterwards, Reichenbach added that, for these reasons, Bohr's "principle of complementarity" acquires profound significance. "When he calls the wave and the particle descriptions complementary, this means that for cases in which one is an adequate interpretation, the other is not." It is a question, therefore, of different descriptions of a certain field of unobservable phenomena.

Finally, Heisenberg states that the law of equivalence between mass and energy is not a reason for despising the older philosophies which upheld the indestructibility of matter. "The terms 'substance' and 'material' of ancient or medieval philosophy cannot be identified with the word 'mass' of modern physics. If we wish to express our actual experience in the language of ancient philosophy, we might consider that mass and energy are two different forms of the same 'substance', thus preserving the idea that substance is indestructible[3]."

143 The Vienna Circle

In Vienna, between the two wars, a small group of philosophers, physicists and mathematicians (Schlick, Franck, Carnap, Hahn, Neurath and others) — known, significantly, as the "Verein Ernst Mach" (Ernest Mach Society) and later generally referred to as the

---

[1] H. Dingle, "The rational and empirical elements in physics", in *Philosophy*, London, **13** (1938), pp. 148–65.
[2] H. Reichenbach, *The Rise of Scientific Philosophy*, Berkeley, University of California Press, 1954, pp. 186, 188.
[3] W. Heisenberg, *Physique et philosophie*, French edition, Albin Michel, Paris, 1961, p. 131.

B

"Vienna Circle" — made a thorough study of Wittgenstein's *Tractatus*. In 1929 they published a manifesto entitled *Wissenschaftliche Weltauffassung* (Scientific Conception of the Universe) which aroused worldwide interest.

The leaders of the movement, while conserving the independence due to each from his position, were at one in rejecting any metaphysical proposition whatsoever, describing them in sweeping terms as "lacking in meaning" and stressing the exclusive value of experience. To a certain extent they were concerned with a revival of old principles — a revival which, because of new justifications and developments, was taking on the name of "logical neopositivism". Both for the philosophers of the Vienna Circle and for Russell and Wittgenstein, logical analysis and syntactic rigour constituted the basic rule of scientific activity. A leading exponent of neopositivism, R. von Mises, has expressed it thus: "All enunciations with meaning can be divided into two groups: those which describe experimental and controllable data, and those which, independently of experience, according to their content, are true or false."

Von Mises, quoting Carnap, develops the argument as follows: "The sequence of words 'Caesar is and' appears to us, according to ordinary grammar, unacceptable because at the place where 'and' stands should stand a predicate and not a connective. The sentence 'Caesar is an emperor' is grammatically correct but a false proposition; the sentence 'Caesar is a prime number' is grammatically correct but meaningless because in a more complete, logical grammar one does not consider the words 'prime number' and 'emperor' merely as substantives but distinguishes between them according to their 'syntactic category'. Then the rule would have to be added that 'prime number' can be predicated only of numbers, 'emperor' only of persons[1]."

1431 "Verifiability" (Schlick) and "connectibility" (von Mises)

The problem of completely eliminating in science the inadequate expression of natural facts suggested to Schlick, the moving spirit of the Vienna Circle, the idea that a proposition can be declared true only when it is "verifiable". This point of view, however, appears unsatisfactory because it would then be necessary to reject as unverifiable all general enunciations, e.g. "the number e is the limit to which the value of the binomial $\left(1+\dfrac{1}{m}\right)^{m}$ tends as m tends to infinity"; "the

---

[1] R. von Mises, *Kleines Lehrbuch des Positivismus*. English version: *Positivism*, Harvard University Press, 1951, p. 72.

English are reserved"; "stray dogs are dangerous"; "the sequoia tree lives three thousand years".

More in keeping with the needs of scientific language, von Mises suggested introducing the word "connectibility" into conventional speech. A proposition is connectible when it appears sensible in the agreed language. For this reason, not even an expression such as "Caesar is a prime number" should be judged *a priori* to be without meaning, but should be evaluated only in the system of ideas and in the context in which it is used. The conventional language of a key to a cipher which happened to designate proper nouns by natural numbers, would make that expression true if, for example, Caesar corresponded to the number 11, and false if he was identified with the number 12, or with no number. The progress achieved by this principle might at first seem insignificant, but it is not so.

The critics of language often quote Hegel's famous formula, *Das reine Sein und das reine Nicht ist doch dasselbe*, and one less known of Heidegger, *Das Nicht selbst vernichtet*, as typical examples of meaningless propositions. Nevertheless, neither Hegel nor Heidegger was mentally deficient, and historians of philosophy understand what they meant.

The proposition of von Mises is very interesting in a wider field because, with the theory of connectibility, it indirectly vindicates the claims of metaphysics which the generality of neopositivists, if not he himself, had condemned in the name of empiricism (section 12).

144   Object language and metalanguage

In order to examine more closely the problems of scientific language, it is advisable to introduce here the strictly indispensable terminology needed in scientific subjects.

Carnap and Morris distinguish the language by which the facts and theories of a science are expounded, from the "metalanguage", which is used when the former is under a critical or interpretative analysis. In the metalanguage, then, scientific expositions become an "object language".

A physicist may say, for example, that the descent of a stone to the ground is the fall of a weight, and one can then explain that the phrase "fall of a weight" is a proposition expressing a principle of physics. (The words and propositions belonging to metalanguage will, as the need arises, be distinguished from scientific expressions by inverted commas.) Similarly, we use the word mathematics to indicate the signs and systems proper to this science, and the word "metamathematics" to describe any observation or description having the mathematical systems as object. The necessity for a metamathematics, proposed and so named by Hilbert, derives from the need to discuss the symbols and expressions of a rigorously formalized mathematics, and to have "a modern instrument for the study of deductive theories".

As we shall see, metalanguage and metamathematics often make use of the forms of common speech, but at other times they create convenient symbols and rules of transformation. This confers on metalanguage a propositional calculus of its own (section 24).

All the metalanguages, then, have a "semantic" function relating to the objects they represent (from *semantikós* = significant); a "pragmatic" function (from *pragmatikós* = practical) concerning the relation between the representation and the interpreter; and a "syntactic" function (from *suntaktikós* = of connection) concerning their internal self-consistency[1].

145   Abstract and concrete disciplines and their language

For disciplines expressed in language, the distinction between "abstract", "analytical" or "formal" subjects and "concrete", "synthetic" or "practical" subjects has a rational basis. They had already been identified in part by Hume when he spoke of "relations of ideas" and "matters of fact" (*A Treatise of Human Nature*, book I, section V). We shall use these terms, however, in a more restrictive sense. Except for what will be said in sections 583 and 587, only logic and mathematics will be named "abstract" subjects, and all the factual or natural sciences will be called "concrete".

---

[1] Many texts could be quoted concerning the ideas in this section, but the following are essential: C. W. Morris, *Signs, Language and Behaviour*, Prentice Hall, New York, 1946; C. W. Morris, "Foundations of the theory of signs", *International Encyclopedia of Unified Science*, Chicago, vol. 1, no. 2; C. W. Morris "Empirismo scientifico" in the volume by various authors entitled *Neopositivismo e unità della scienza* (see above); R. Carnap, "I fondamenti logici dell'unità della scienza", in the volume *Neopositivismo*, etc. See further A. Pasquinelli, *Linguaggio, scienza e filosofia*, Il Mulino, 1961; *I problemi del linguaggio*, Accademia dei Lincei, Rome, 1962, which contains the transactions of a conference in 1956, with important reports by L. Geymonat, M. Fubini, L. Ronga, L. Venturi, C. L. Musatti and B. A. Terracini, and comments from eminent people (in particular F. Severi, A. Pagliaro, G. Bonfante and C. L. Ragghianti, etc.). Finally, specifically interesting and with an extensive bibliography, the essay by L. Geymonat, *La metamatematica dopo Hilbert*, in *Atti del VII Congresso dell'Unione Matematica Italiana, 1963*, Cremonese, Rome, 1965; and J. Jörgensen, "Origini e sviluppi dell'empirismo logico" in the volume *Neopositivismo*, etc.

It is not necessary here to consider whether those branches of philosophy which are certainly not identified with logic should be assigned to the abstract sciences. Neither do we insist on the differences between factual sciences and historical disciplines. Apropos of these last, we shall merely say that the narratives of history, with their precise singling out of events (the deeds of Alexander the Great and of Louis XIV, and not the things that happen to kings; the feats of arms at Cannae and Austerlitz, and not battles in general), and the appreciation of art (the Hermes of Praxiteles, and not ancient statuary; Wells Cathedral, and not Gothic architecture; Beethoven's Ninth Symphony, and not choral orchestral compositions), differ profoundly from statements such as those relating to definite proportions in chemistry, stratification in geology, the sex ratio in populations, consonantal shifts in linguistics, and so on — these latter being all phenomena described by the various sciences in terms of generalization, i.e. deprived of any sign of individuality.

15 Frege, Peano and Russell on logic, arithmetic and mathematics

The language problems in the two categories of the sciences have their own peculiarities and therefore deserve to be considered separately.

As regards the abstract disciplines, Russell comments as follows: "Frege was the first to expound the idea (1884), which was also mine and which I still uphold, that mathematics is an extension of logic, and he was the first to define numbers in logical terms." This same idea, which Peano had expressed almost at the same time as Frege, was developed by Russell in collaboration with A. N. Whitehead in a now famous work entitled *Principia Mathematica* (3 vol., 1910–13). The treatment was not completely accepted by Hilbert. Nowadays it has lost its revolutionary flavour and has brought into use the idea of the sequence: logic → theory of numbers → mathematical sciences[1].

The connection between arithmetic and logic is clearly seen by noting that Peano's arithmetic was based on three fundamental ideas:

zero, number (i.e. whole number), successor,

and on five logical principles:

(1) zero is a number;

---

[1] B. Russell, *Introduction to Mathematical Philosophy*; B. Russell, *History of Western Philosophy*, Allen and Unwin, 1945; L. Geymonat, "Giuseppe Peano", in *La Scuola in Azione*, 1959–60, No. 7; A. N. Whitehead, *Science and the Modern World*, 1925.

(2) the successor of any number is a number;

(3) no two numbers have the same successor;

(4) zero is not the successor of any number;

(5) any property which belongs to zero, and also to the successor of every number which has the property, belongs to all numbers[1].

Some of these principles will be referred to later.

151   Wittgenstein on the relation between logic and mathematics

Wittgenstein put great emphasis on the sequence connecting mathematics to logic, as witness the following selection of axioms from the *Tractatus*:

5   A proposition is a truth-function of elementary propositions.

5.43   All the propositions of logic say the same thing, to wit, nothing.

6.1   The propositions of logic are tautologies.

6.11   Therefore the propositions of logic say nothing (they are analytical propositions).

6.1262 The proof in logic is only a mechanical expedient to facilitate the recognition of tautologies in complicated cases.

6.2   Mathematics is a logical method.

6.21   A proposition of mathematics does not express a thought.

6.3   The exploration of logic means the exploration of *everything* that is subject to law. And outside logic everything is accidental.

Now, whereas Peano stated the foundations of arithmetic in a purely logical form, Wittgenstein affirms that mathematics can be reduced to logic, that logic and mathematics do not have a factual content, that logical and mathematical proofs are only analyses of the premises, and that outside logic and mathematics — the domain of formal necessity — everything is accidental. Wittgenstein is insistent about this idea and says: "No necessity exists by force of which a thing must happen because another thing has happened. Only a logical necessity exists" (6.37). And earlier, referring to a statement of Hume's, he had written: "That the sun will rise tomorrow is a hypothesis: and this means that we do not 'know' whether it will rise" (6.36311).

Considered attentively, axiom 6.37 is clear. Wittgenstein denies, as does Hume, the idea of causality. Concerning the past, he recognizes

---

[1] B. Russell, *Introduction to Mathematical Philosophy*, p. 6; L. Geymonat, "Giuseppe Peano"; L. Geymonat, *Filosofia e filosofia della scienza*, 1962.

the historical sequence of certain events, while as regards the future he does not affirm that one cannot make forecasts of occurrences in Nature, but he maintains that no logical argument could justify them. He also, like Hume, knew that the sun would rise on the morrow, but neither he nor others were in a position to demonstrate it (sections 31 *et seq.*, 3322).

152 Tautology of logical and mathematical proof

Pausing for a moment on the subject of mathematics: even a beginner understands that its principles are actually developed without adding anything to what its premisses contain in essence — that is, tautologically (from Greek *tautó* = the same). To clarify the ideas, consider the very simple theorem about the sum of the internal angles of a triangle. The elementary demonstration of it produces a momentary satisfaction, but this vanishes as soon as one understands that the sum being equal to two right angles is derived directly, in Euclidean geometry, from the postulate of parallels and is part of the very definition of a triangle.

But in that case, of what use are proofs? Wittgenstein replies aptly with proposition 6.1212: they render explicit that which the definitions contain implicitly and which sometimes requires long and difficult reasoning to make clear. Absurdly enough, this seems confirmed by the fact that there are theorems almost certainly true, which no-one has been able to prove. Such is the so-called last theorem of Fermat: there do not exist solutions in whole numbers of the equation $x^n + y^n = a^n$, where n is greater than 2. That a proof exists is not to be doubted in view of a note left by Fermat in a book he had been reading: "I have discovered a really beautiful demonstration, which this margin is too small to contain". But, unlike many theorems which he left undemonstrated and which were subsequently proved by others, his last statement still remains a mystery.

Other theorems, although true, are by their very nature undemonstrable. Russell proposed the following: "All the prime numbers which no-one will have thought of before A.D. 2000 are greater than one million". Apart from being undemonstrable, this obvious theorem cannot be exemplified.

153 Justification of the premisses of logic and mathematics

But, at first sight, a justification of Wittgenstein's categorical assertions is not easy — namely that the propositions of logic do not say anything; that the propositions of mathematics do not express any thought, they are tautological. Clearly a purely literal interpretation would lead to the conclusion that logicians and mathematicians are simply idlers. This was not what the author intended; he simply meant to say that logic and mathematics are a formal development of the premisses, without factual content. But then the question is, are the premisses part of logic and mathematics, and are these disciplines empty of all content? The affirmative answer which must be given to the first question — since it is the only one which justifies the "empty" developments — restores to the disciplines which are derived from the premisses a content for the very purpose of which these premisses were formulated. Perhaps the motivation was based on some remote factual suggestion, or — according to Hilbert — on intuition, or on some psychological need to propose a problem and to set oneself to resolving it. These motives might determine the content of the logical and mathematical premisses, and therefore of the developments from them, which could conventionally be called tautological, indicating that they were concerned with no other external data beyond the personality of whoever was occupied with them.

16   The language of mathematics

Mathematics has long since discarded the ordinary discursive form of expression in favour of a symbolism with its own strict rules of syntax. These set the reasoning on an automatic course and reduce error to the minimum. We must emphasize the effect which mathematical language has exerted, and still exerts, on the factual sciences; and we recall (referring back to section 142) that certain of these, such as chemistry and to some extent genetics, have made notable progress in the use of formalized expressions. We leave out of consideration for the moment many natural sciences which make use of mathematics in their progress from qualitative observation to quantitative expression of the physical world.

161   Suitability of mathematical language for logic

The subject which, by feedback, has received the greatest impulse to adopt the symbolism of mathematics is logic, and "Logistics" is the modern name for formal logic[1]. Since symbolic reasoning in logic is less familiar than in mathematics, we must consider it in some detail.

---

[1] *Translator's note* — In English, "logistics" also has the meaning of the provision of supplies, especially in a military context. The adjective has also a specialized meaning in statistics, as relating to a particular kind of growth curve.

Short of tracing the movement towards the formalization of logic back to Aristotle, as Scholz maintains, we may say that it has its beginning and development in the flowering of Renaissance science. We have referred to the importance of technical language in the natural sciences, pointing out that it appeared with the great discoveries of Galileo and with the revolution in chemistry introduced by Lavoisier.

A rudimentary formalization in logic is very old. It is already clear in the first figure of the syllogism "If every A is B and C is A, then C is B" (see section 1631); and it is evident that the clarity of such formulae facilitated the rediscovery of all the 64 modes in which the syllogism used to be classified. But the later Renaissance requirements were on a wider scale. During the seventeenth century three personalities of unequal stature, but all with penetrating insight, pointed out within a few years of one another the need for formalizing expression: René Descartes (1596–1650), the Very Reverend John Wilkins (1614–1672) and Gottfried Wilhelm Leibnitz (1646–1716).

1611 Descartes' contribution to the formalization of philosophic language

In a letter to Father Mersenne[1], dated November 20th, 1629, after making fun of the inventor of an artificial language who was praising it to blow his own trumpet, Descartes presents a brief outline of a philosophical language. In this, thoughts were to be classified in such a way as to be rapidly learnt, "as one can, in a day, learn to name all the numbers up to infinity and to write them in a strange language, although there may be an infinity of different words". "The invention of this language," he adds, "depends on true philosophy, since it is impossible otherwise to enumerate all the thoughts of men and to put them in order, or even to distinguish them so that they are clear and simple." It would be "a very easy language to learn, pronounce and write, suitable for representing things so distinctly that it would be impossible to go wrong. . . . I am certain that this language is possible."

1612 Wilkins' contribution to the formalization of scientific language

The Very Reverend John Wilkins, Bishop of Chester and, for a time, President of the Royal Society, did indeed construct such a language, and after years of work he expounded it in a book published

---

[1] R. Descartes, *Oeuvres et lettres*, Bibliothèque de la Pléiade, Paris, 1952, p. 911.

B*

in 1668[1]. Wilkins' written language was not an easy one — although this was the aim in view — as the words were represented by signs similar to Chinese ideograms. All knowledge was catalogued hierarchically in classes and in various subdivisions, to which the signs corresponded. These were connected by prefixes, suffixes and other symbols, according to the rules of a logical grammar.

Wilkins' basic error is apparent from this brief mention. Instead of looking for a way to express logical syntax, he became involved in the variable semantic content of scientific discourse (section 144), so complicating his laborious work as to render it useless.

1613 Leibniz, true originator of the formalization of logic

Descartes had outlined the project for a formal philosophic language and Wilkins constructed a scientific one, but the most promising way, that of the formalization of logic, was indicated by Leibniz. Modern authors credit him with stating, briefly and impressively, the criteria which were later developed into logistics.

Lewis and Langford, in the historical introduction to their classical treatise on symbolic logic[2], note that Leibniz had a Utopian scheme for the reorganization of knowledge by introducing a scientific language (*characteristica universalis*) to be worked out algebraically (*calculus ratiocinator*). They deplore "the exaggeration of the role attributed to deduction as against induction, because of the rationalistic point of view". But they add: "The *calculus ratiocinator* was a more limited project. Had it been completed, it would have coincided with what we call today symbolic logic. That is, it would have been an instrument for reasoning in general, developed in ideographic symbols for working out logical operations. Here Leibniz achieved a certain degree of success. . . . However, the foundations he used were not such as could be built on, since he did not succeed in freeing himself from certain traditional premises or in solving the difficulties caused by them."

Evaluating him more favourably, Scholz represents Leibniz as the only logician worthy of comparison with Aristotle. He credits him with

---

[1] See the short article by M. Boldrini, "Alle origini della logica simbolica", in a collection of *Studi in onore di A. Fanfani*, Giuffrè, Milan, 1962.

[2] C. I. Lewis and C. H. Langford, *Symbolic Logic*, 2nd edn, Dover Publications, New York, 1959. There are also the following important Italian works on symbolic logic: A. Pasquinelli, *Introduzione alla logica simbolica*, Einaudi, Turin, 1957; E. Casari, "Computabilità e ricorsività, problemi di logica matematica", in *Quaderni della Scuola di Studi Superiori sugli Idrocarburi*, 1959; and E. Casari, *Lineamenti di logica matematica*, Feltrinelli, Milan, 1959.

the discovery that deductive syntax is independent of semantic content (section 161), and also with the discovery of the possibility of a calculus of logical propositions which does not resort to quantifying the qualitative. On the contrary, he proposes (on the basis of recognizing mathematics as a branch of logic) "a large-scale requalifying of the quantitative". Only since the publication of Russell's *Principia*, says Scholz, can logic come into possession of an instrument suitable for developing the Leibnizian premisses[1].

The following fragment, quoted by Pasquinelli, brings out forcibly the breadth of Leibniz's horizon. "Id ... efficiendum est, ut omnis paralogismus nihil aliud sit quam error calculi. . . . Quo facto, quando orientur controversiae, non magis disputatione opus erit inter duos philosophos, quam inter duos computistas. Sufficiet enim calamos in manus sumere sedereque ad abacos, et sibi mutuo ... dicere: 'calculemus'."[2]

162 Boolean algebra

It is only with certain mathematicians of the first half of the nineteenth century, notably A. de Morgan and above all G. Boole, that we find a true system of symbolization of logic, that is, a written form suitable for the automatic elaboration, not of relations of quantity but of propositional functions. In the logical algebra of Boole (already formulated in 1847 and expounded later in the better-known work of 1854) is the following theorem, which gives a first idea of logistic reasoning[3].

If x is a class (e.g. the class of all sheep which have horns), the operation of "electing" them twice does not change the resulting class. In symbols this is correctly expressed thus: $x \times x = x$ or otherwise by $x^2 = x$, a formula which seems absurd to anyone thinking in algebraic terms.

Among Boole's merits is the lasting one of having shown that his logical algebra also becomes valid mathematically when the class

---

[1] H. Scholz, *Storia della logica*, pp. 101–2. There are numerous quotations from Leibniz and a lively reconstruction of the latter's contributions to logistics.

[2] G. W. Leibniz, *Die philosophische Schriften*, Berlin, 1875, vol. VII, p. 200, quoted by A. Pasquinelli, *Linguaggio, scienza e filosofia*, p. 71.

[3] George Boole, *An Investigation of the Laws of Thought, on which are founded the Mathematical Theories of Logic and Probability*, 1854, since reprinted by Dover Publications, New York. See pp. 30 *et seq.*, also C. I. Lewis and C. H. Langford, *op. cit.*, p. 13.

symbols (election symbols) x, y, ... are translated into quantities and limited to the values 0 and 1, to be understood as "absence" and "presence" or also as "none" and "all". In fact, in the case mentioned above, $0^2 = 0$ and $1^2 = 1$. In these concepts we are first introduced to binary arithmetic which nowadays has taken on such importance in electronic calculating (see section 23).

163   Logistics, symbols, and propositions

Modern symbolic logic or "logistics" is so highly technical that it requires considerable study, such as is needed to understand ordinary algebra.

Logical algorithms, like those of mathematics, consist of symbols and formulae. A list of the basic symbols with their meanings is given below:

Sign	Gödel's elementary numbers	Meaning
~	1	not
v	2	or
⊃	3	implies
∃	4	there is a ... (universal quantifier)
=	5	equals
0	6	zero
s	7	the successor of
(	8	punctuation mark
)	9	punctuation mark
,	10	punctuation mark

Various authors introduce other signs such as "." for the conjunction "and", "$\neq$" for inequality, "$-$" for subtraction, and x, y, z as variables.

On the other hand, not all the additional symbols are indispensable to the formalized language of logic and metamathematics. Thus, for example, the sign of conjunction "." corresponds to the negation "~" of two negations, viz.

$$(x.y) = {\sim} ({\sim} x \ v \ {\sim} y)$$

In its turn, the sign "$\neq$" clearly corresponds to the negation of "$=$".

Using the previous symbolism, a proposition such as "every number has a successor" (which is the second of Peano's postulates — cf. section 15), the formal definition of which would be "there is an x such that x is the immediate successor of y", assumes the following form:

$$(\exists x) (x = sy)$$

With suitable rules, Gödel's numbers could also be substituted for the above symbols. We shall go into this later.

Four basic propositional relations are given by Lewis and Langford, and the following have been adapted to the preceding table and constitute the elements of syllogistic reasoning. In the first column the relations are shown in ordinary language, and in the succeeding column are given the forms which they take in symbolic logic[1].

(1) All a's are b's: $a \supset b; a.b = a; a. \sim b = 0$
(2) No a is b: $a \supset \sim b; a. \sim b = a; a.b = 0$
(3) Some a's are b's: $a. \sim b \neq a$
(4) Some a's are not b's: $a.b \neq a$

These are the propositions: universal affirmative, universal negative, particular affirmative, and particular negative. It can immediately be seen that proposition (1) contains in the form $a.b = a$ the basic theorem of Boolean logic.

1631 Logical statements and propositional calculus

Propositional calculus is too complex a subject to have only passing mention. One must, however, insist on the formal character of logic, in that it leaves out of consideration any judgement on the validity of the facts concerned.

A logical statement consists of propositions with variable parts conjoined by invariable parts, and has the following schematic form, as mentioned in section 161.

If every ... is ... and ... is ... then ... is ...

This always becomes valid when the six empty spaces between the invariable parts are filled by the same word in spaces one and four, two and last, three and five. For instance, the example given in every textbook becomes: If all men are mortal and Socrates is a man, then Socrates is mortal. But it should be stressed that the syllogism is valid only when the spaces in the formula are filled in according to the rule.

Anyone who said "If every cock is Aesculapius and Crito is a cock, then Crito is Aesculapius," while making a false combination (and in bad taste) of the last words of Socrates, would be making a logically

---

[1] The algebraic sign $<$ would suggest the form $\subset$ (instead of the traditional $\supset$) to express implication, but for this latter, modern usage prefers the sign $\rightarrow$. According to other writers, the sign $\supset$ is adopted when a and b are used as predicates, while when a and b are used as classes the written forms $\supset$ or $\subset$ are optional.

correct statement, given that logic does not enter into the meaning of the propositions. Copi presents another argument, less far-fetched, but equally valid in form while false in substance: "If all spiders have six feet and hexapods have two wings, then all spiders have wings."

These observations allow of a distinction between "propositions", the truth of which is subject to factual criticism (semantic), and "arguments" or "statements" whose validity (syntactics) depends entirely on keeping to formal rules.

In the preceding syllogism there are three propositions of which the premiss — all spiders have six feet — and the conclusion — spiders have wings — are in conflict with scientific truth, while the second — hexapods have two wings (for this reason they are also called Diptera) — is accurate. Because the formal rules are respected, the entire argument is logically correct.

These considerations complete what was said about Wittgenstein's axiom, i.e. that the propositions of logic say nothing. The meaning of this axiom has already been made clear with regard to the abstract sciences (sections 151 and 152), but the axiom remains true whether the propositions composing the argument are abstract or concrete.

To complete these ideas, we give Pasquinelli's transcription in symbols of the general statement of the syllogism discussed above[1]:

$$[(x)(Q(x) \supset P(x)).(\exists x)Q(x)] \supset (\exists x)P(x)$$

17   The empirical sciences regarded as language by Carnap

Once the abstract character of logic and mathematics is fully confirmed, the conclusion follows easily that they can be thought of as purely linguistic systems.

But how can one accept the view of the neo-positivist philosophers who affirm (with an obvious contradiction in terms) that not only the abstract sciences but also the so-called empirical ones are purely a matter of language? One can agree with the objection that a large part of what science propounds depends on the medium of expression which it uses. But there is a great difference between such an attitude and the radical position of Carnap, who establishes that the terms needed to define the data and processes of inorganic matter are "physical terms", and that "physical language" is the scientific sub-language which, apart from mathematical and logical terms, contains only

---

[1] A. Pasquinelli, *Introduzione alla logica simbolica*, pp. 19 and 54; I. M. Copi, *Introduction to Logic*, p. 10. See also: W. van Orman Quine, *Elementary Logic*, Boston, 1941.

physical terms. He continues: "The system of those statements which are formulated in physical language and are acknowledged by a certain group at a certain time is called the physics of that group at that time. Such of these statements as have a specific universal form are known as *physical laws*." Carnap defines biology and its laws analogously[1].

171 Mach's "elements" and the opinions of Dingle and Popper

Yet the conflict is more apparent than real. Some years ago Dingle emphasized more precisely than anyone else had done this simple truth: that science is always a mental construct, by means of which a collection of objective data is arranged in a model and expressed linguistically for certain ends[2]. Furthermore, this is true not only for major concepts such as the theory of relativity, quantum mechanics, the theory of economic equilibrium and so on, but for the generality of so-called factual data, including those which are expressed by one word, when this word synthesizes a more or less complex group of sensations. According to Mach's theory of "elements"[3] (to which we shall return in sections 172 and 311), only certain direct expressions of the senses and the words which express them have an elementary character, such as a colour, a sound, a temperature, a space, or a painful sensation, while all their combinations which form objects, places, feelings and so on, have an abstract character. A chair, a table, toothache are not elementary sensations but "ideas"; and precisely for this reason an ordinary table can be described in the various ways mentioned in section 11.

Popper, a neo-positivist philosopher[4], theorizing on the problems of the social sciences, points out that abstractions, and therefore words, form the greater part of the objects with which these sciences are concerned. It may appear strange to some people, he says, that "war" or

---

[1] R. Carnap, "Logical foundations of the unity of science", in the volume by various authors, *International Encyclopedia of Unified Science*, vol. 1, p. 46.

[2] H. Dingle, *op. cit.* For the Italian edition of this work the author added an explanatory note to the effect that the thesis of his study, written a quarter of a century before, was still valid, although, at the time of writing, documentation of the physical sciences had a decided advantage over the intellectual arrangement of data in 1938, when relativity and quantum theory were under lively discussion, and this systematization seemed to absorb the greater part of scientists' attention. Concerning the alternation of active development and periods of reflection in the sciences, see L. Geymonat, *Filosofia e filosofia della scienza*, p. 80.

[3] E. Mach, *op. cit.*

[4] K. R. Popper, *The Poverty of Historicism*, 2nd edn, *Routledge & Kegan Paul*, London, 1960.

"army" are also abstract concepts, while the concrete reality would be rather "men killed" and "men in uniform". The example is not a very good one, but it indicates the still greater complexity of Popper's final expressions, which are still ideas and not Mach's "elements".

Dingle considers the model of the electron and says that it includes factual data — e.g. the greenish, luminescent trajectory seen and photographed in a Wilson chamber — and a considerable number of concepts. The combination of some of these leads to the corpuscular conception, and of others to the wave interpretation of electrons. Therefore electrons are not a fact but a word in the language of subatomic physics. Clearly, all discussions in which this word is used draw their validity from their factual consistency and from the non-contradiction of their logical system of expression.

172 Mach's "elements" and von Mises' "protocol sentences"

Mach has been mentioned several times as a thinker who, with the theory of "elementary sensations" or "elements", helps to elucidate the link between objective knowledge and language. When someone says "I see the colour red" or "I feel cold" he is expressing a fact of which he is immediately conscious and which cannot be reduced to anything more elementary and correct. But if he says "the red robe of the cardinal has a train" or "this room is overheated", the two expressions imply a superposing of concepts formed from sensations and from their intellectual combination.

Consider now the distinction proposed by Carnap between "ways of speaking by content", that is, presented as if they set out a simple factual state, and "formal ways of speaking", in which one is concerned with strict syntax and hence with keeping distinct the factual and non-factual elements of the proposition. It is immediately clear that words and propositions concerning Mach's "elements" are factual and irreducible, while one can at least consider the possibility of formulating complex propositions in a syntactically reliable way, apart from the semantic aspect.

The exponents of logical empiricism call very simple propositions which describe a sensation "atomic" or "protocol sentences", thus alluding to their irreducibility and comparing them with the first rough notes that a research worker enters up.

Von Mises was the first to draw attention again to Mach's elements, considering them as an essential part of atomic propositions[1]. Although

---

[1] R. von Mises, op. cit., pp. 82 et passim.

his theory is none too clear, he seems to maintain that complex propositions of content, when they are formed of elements — or, if preferred, of protocol propositions — can always be translated into formal propositions of the kind which some exponents of empirical logic call "molecular". Thus the basic statement of Mach's theory, says von Mises, is the following: "For us the world is composed of a complex of sensations". In formal terms this becomes: "Our statements, sentences and questions about the world are reducible to statements, sentences and questions about sensations". And more particularly the words "A body is a complex of sensations of a special kind" become the following molecular proposition: "Every sentence containing the word 'body' is equal in content to a system of sentences in which the word 'body' does not occur, but in its place there is reference to the elements in certain relations".

One finds oneself, as can be seen, in the narrow orbit of a syntactical analysis.

173 Obstacles to the general use of symbolic language in science

Many exponents of logical empiricism have thoroughly worked out the problems involved in the use of logical symbolism in the treatment of scientific questions. Carnap, in particular, habitually uses symbolic logic in his studies; for example, all the theorems of his famous treatise on probability are written in logistic terms[1]. However, he is well aware that symbolic logic is insufficiently developed and that its chances of coming into general use are slight. At present the idea of making it the usual instrument of scientific expression is unrealizable. As Jörgensen[2] points out, Carnap often translates into words the symbols he uses, even in his theoretical studies. Other logicians, such as Harrod[3], definitely reject logical symbolism, being satisfied with the strictness obtainable by the use of ordinary language.

Apart from specialized cases, logical symbolism is in fact making slow progress. It is not systematically used in this book, for practical reasons which will be explained in section 182. However, logical formalization has good prospects of making headway; a hopeful sign has been the courageous initiative of a progressive lawyer, C. Magni,

---

[1] R. Carnap, *Logical Foundations of Probability*, 2nd edn, University of Chicago Press, 1962.

[2] J. Jörgensen, *Origini e sviluppi dell'empirismo logico*, p. 172.

[3] R. F. Harrod, *Foundations of Inductive Logic*, Macmillan, London, 1956.

who included a dissertation on symbolic logic as a preface to a substantial work of his[1].

18   Applied and pure mathematics

A question that has had insufficient attention, despite its paramount importance for science, concerns that special language known as "applied mathematics". The idea tends to be accepted as unconditionally valid that pure mathematics and scientific mathematics are the same thing, and that consequently the latter needs no adjustment when the algorithms of the former are used in studying natural phenomena. This is true only up to a point, however, and at the root of the idea there still lies the obsolete Galilean assumption of a Nature ruled by mathematical laws, which the physicist must limit himself to discovering and expressing in formulae. The subject is of such vastness as to require more than a passing reference at the end of a discussion devoted to the problems of scientific language; but the references already made to the mutability of scientific laws (section 112) should suffice to make it clear that nowadays the positions are reversed. It is not that Nature's laws are within human reach, but rather that the mind must adapt mathematical instruments to its provisional picture of the world in order to simplify it, make it intelligible, summarize its salient features, and work out unforeseen regularities and new groupings of phenomena. Moreover, in the recent past, the relations found in Nature were thought of as necessary and eternal, thereby lending themselves to theoretical analysis. But a more thorough probing of reality and a sharper criticism of concepts previously suggested led, after long consideration, to the acceptance of a new branch of mathematics, that of causal relations. It made evident the radical discontinuity of the subatomic world, a discontinuity which is ignored in certain cases, but only so as to simplify analytical language.

These problems really merit a much more thorough treatment. From the foregoing it will be clear that the mathematics of science is not pure mathematics but a complex of ideas and techniques brought into the service of practical disciplines and sometimes improved and developed through them. In particular, the mathematics of static and dynamic probability (stochastic processes) and that of discontinuous variables (equations in finite differences) are original creations of science, even if pure mathematicians have contributed to their refine-

---

[1] C. Magni, *Avviamento allo studio analitico del diritto ecclesiastico*, Giuffrè, Milan, 1956.

ment. Mathematics has put at the service of the sciences an immense, traditional, and valid instrumentation.

181 Numbers in science as used for quantifying

Numbers, although certainly originating in experience (as is confirmed by the universal use of the word "digit" for number, from Greek and Latin down to modern languages), are thought of by philosophers as a direct achievement of logic, and the entire body of mathematics has developed from them.

Now numbers are an indispensable tool of scientific research. In the natural sciences everything is quantity, as are also verbal concepts of measure such as large and small, first and after, dear and cheap, and so on. For this reason, science relies increasingly on mathematics and measurement.

But numbers in science only formally tally with those in mathematics. A trained mind has no difficulty in appreciating the difference between, let us say, abstract numbers such as e and π and natural constants such as the acceleration of gravity or Planck's constant h. In the first case we are dealing with quantities expressing relations or mathematical properties, in the second with measures of events or things.

The subject of quantification in science also merits separate discussion (section 46), but here it is enough to emphasize the linguistic nature of numbers which, though originating in pure mathematics, have been taken over by science.

So far as science is concerned, the definition which logicians give of number is quite unsatisfactory, e.g. "that thing which is the number of a given class" (Russell). In its place should be put an idea which recognizes the need of being adaptable from time to time ("concordable" in von Mises' sense) to the language and propositions peculiar to the branch of knowledge in which numbers are used.

1811 Buffon and Condorcet on applied mathematics

Although anticipating certain ideas not yet introduced into this study, this seems a suitable place to quote an extract from a classical author, who outlines many of the questions hitherto discussed. In the eulogy which Condorcet wrote about Buffon, after his death, he refers to the latter's Essay on Moral Arithmetic in the following comments.

"He puts forward the opinion that mathematical truths are not real truths but simply truths by definition [today we should call them tautologies]. Metaphysically this is correct, but it applies equally well to truths of every kind, so long as they are precise and do not have individuals as their subject (i.e. they refer only to defined groups).

"If, then, we wish to apply these truths in practice and thus render them specific to individuals, they become no more than approximate truths.

"There really exists only one difference, which is that the ideas which form mathematical or physical truths are more abstract in the former. Hence it follows that for physical truths we have a clear knowledge of certain individuals to whom they apply, but no knowledge of other individuals. But true reality, the usefulness of any general proposition, is independent of individual experience. A statement which applies to every case is in fact an absolute truth or approaches indefinitely near to truth".

In this passage what really matters is the emphasis placed on the tautological nature of mathematical reasoning and the special nature of applied mathematics. Buffon is also concerned with the separation of content and form (an Aristotelian idea), and he hints at the propositional (linguistic) character of that scientific reality which can be expressed mathematically.

182 Instrumental character of applied mathematics

Since "applied mathematics" is the usual language of empirical subjects, one understands why it has neither thoroughly developed, nor generally adopted, a logistic language. An "applied mathematics" language might have improved the process of reasoning, but it would have obscured and limited communication between the scientists who originated it and those who used it. Thus one understands why the common tongue is still appropriate to the scientific character of this book, despite the epistemological problems that face us[1].

Once mathematics was recognized as the chief scientific language, the old intractable problem of explaining how natural facts can be formalized fell away, almost as if Nature were really written in that mathematical language about which Galileo enthused. Since science describes not elementary facts but their intellectual arrangement, this latter evidently adapts itself to the formulae (and vice versa) and relies on them for a formal arrangement and a simplifying synthesis. Nobody

---

[1] See the book by S. S. Acquaviva, *Il problema della logica nelle scienze umane*, Marsilio, Padua, 1964, in which logical language is used in formalizing sociology.

has ever been astonished that "butter" is a word that belongs to butter, and nobody should marvel because the conceptual structure of, for example, the phenomena of interference has found expression in the formulae of Heisenberg, or because errors of observation follow a Gaussian distribution.

These cases are quite different from those where certain constants appear in pure mathematics, such as e in expansions in series and π in trigonometric functions. Such constants, being latent in the premisses, become a new concept once they are known and interpreted. On the other hand, the formulae of applied mathematics are the means by which events are simplified and described; these formulae are developed deductively so as to draw from them corollaries for comparing with reality. Sometimes they are transferred by analogy to other fields. A practical example of this process is the analogue computer, in which complex problems of mechanics and technology are simulated.

2

Axioms

21   Kantian *a priori* synthesis

When Kant proudly compared his own discoveries to the Copernican revolution he was certainly not overprizing them. Beginning with penetrating criticisms of empiricism, which seeks to found knowledge on generalization from perceptual data (Locke, Hume), and of rationalism, which postulates the conformity of thought to the structure of the world (innateness of Descartes, monadism of Leibniz), Kant arrived at the new idea of "synthetic judgement *a priori*". This avoided the double error of imagining a natural order independent in itself, and a mind able to free itself from its environment when considering the essence of that order.

"Knowledge", as an authoritative neo-Kantian philosopher has explained[1], "is synthesis *a priori*: synthesis, because it is organization or connection of sensory data; *a priori*, because the beginning of this organization is our consciousness. In this process of organization, it acts according to its own inherent laws, universal and necessary and therefore independent of experience and anterior to it. Science consists of *a priori* synthetic judgements: of judgements, that is, which are extensions of knowledge as empiricism requires (and not simply explanatory, as are the analytical judgements of the rationalists), and, in so far as they are *a priori*, possess that universality and necessity which was lacking in the synthetic judgements of the empiricists.

"Only, therefore, to the extent to which the mind directs sensory material does it transform it into experience, i.e. into natural reality. It is our own intellect which prescribes laws to nature, and not vice versa."

---

[1] E. P. Lamanna, *Storia della filosofia*, Le Monnier, Florence, 1952, vol. II, p. 363.

211 The innovating function of Kantian synthesis

Speaking, then, of his own revolution, Kant definitely knew that he had established that rational basis of the system of ideas which Plato had lacked, once the scepticism of the empiricists and the dogmatism of the rationalists had been overcome and the universality of geometry linked with the reality of the physical world. Hence the insistence with which, in the first of his celebrated *Critiques*[1], he returns to the *a priori* synthetic character of mathematical axioms and contrasts them with the generality of the judgements of the empirical sciences.

He wrote: "Geometrical propositions are all apodictic, that is, needing no demonstration . . . but they can be neither empirical nor judgements from experience, nor be treated as consequences of experience." And again: "The mathematics of spatial extension [geometry], with its axioms, is based on successive syntheses of the creative imagination [meaning the intuitive feeling of space-time] in the formation of figures. These axioms express the conditions of *a priori* sensory vision, under which conditions, however, is realized the schema of a pure concept of the facts of the external world."

212 Post-Kantian revision of *a priori* synthesis

Since 1781, however, much water has flowed under Euler's proverbial topological bridges at Königsberg; many doubts, some constructive, have been expressed about logical *a priori* synthesis; mathematics and science have taken on new viewpoints, thereby imposing a revision of philosophic positions which, starting from the Kantian premises, had become hopelessly isolated.

Nothing could better measure the separation between ancient and modern epistemology than a thorough examination of the new views about the principles of mathematics, with the related ideas of space and the present-day conception of time. But the analysis would go beyond the limits and scope of this study, for which it is enough to dwell somewhat on the old and new meanings attached to the word "axiom".

22 The nature of axioms

All mathematical and scientific literature, from Euclid's geometry to Newton's *Principia* and the moderns down to Kant and later, had taught that there are two categories of *a priori* propositions, namely axioms and definitions.

---

[1] Kant, *Kritik der reinen Vernunft; Tranzendentale Elementarlehre*, 2er Teil.

Axioms, as is well known, dogmatically assert a principle, a circumstance, or a condition. Here are examples:

"We think of three different systems of things, ... points, ... straight lines, ... flat surfaces" (Hilbert); "the world is all that is the case"; "what is the case — a fact — is the existence of states of affairs" (Wittgenstein).

Definitions, on the other hand, are propositions in which, by means of a conjunction (copula), the identity of two linguistic statements is affirmed, such as "a straight line is the shortest distance between two points" (Archimedes); "mass is the quantity of matter" (Newton).

Modern logicians have recognized the tautological nature of definitions in that they can be reduced to the repetition of the same axiom differently formulated, and are thus shown to be a useless traditional left-over. This becomes clear when the preceding definition of Newton's is formulated as follows: "I think of a mass, which I also call a quantity of matter".

The preceding conclusion was clearly expressed in 1883 by Mach and is no longer matter for discussion. In the light of this, von Mises has shown that, in the first propositions of Newton's *Principia*, the basic concepts of mass and force are first defined, and then — on account of the evident impossibility of otherwise explaining the terms in conjunction — they are re-proposed in axiomatic form[1].

221   Rejection of the *a priori* validity of axioms

But during the past century or more (as mentioned above) the idea of axiom itself has also been revised, depriving it of that *a priori* character accorded to it by Kant. This revision has spread from the field of pure mathematics to that of experimental subjects.

In the transition two phases can be recognized, though in part they intersect and overlap. They are: the questioning of the *a priori* nature of mathematical axioms and therefore (for the reasons explained in sections 15 and 151) of logical axioms; and the assimilation to them of certain generalizations from the natural sciences.

222   Empirical origins of mathematical axioms

There had, in fact, for some time been reason to doubt the structural universality of mathematical axioms (section 181). All writers concerned with the subject have pointed out that the first ideas on geometry

---

[1] R. von Mises, *Manuale di critica scientifica e filosofica*, Longanesi, Milan, 1950, p. 148.

were developed in Egypt, from the need to measure landed property and identify boundaries periodically washed out by the Nile floods. The ancient Egyptians learned much from this practical agronomy, as for instance, that a triangle is right-angled if the length of its sides is proportional to the numbers 3, 4, 5. But the logical and inventive Greek mind was needed to place the empirical ideas of geometry on a universal level. Among other things, this enabled Pythagoras to find the general demonstration of the theorem of the right-angled triangle[1].

But geometry (recalling sections 15 and 181) has its origin, like every other branch of mathematics, in the theory of numbers, which in turn is derived from logic, and for this reason numbers — either evident or hidden — appear both in logical formulations and in mathematics. Thus, among the geometrical and logical axioms referred to above, Hilbert starts by recognizing "three" systems of things, Wittgenstein establishes that the world is "all" that is the case, and Archimedes defines a straight line as the "least distance" between "two" points.

Conversely, however, the reduction of mathematics to logic has determined the development of that algebra of qualities of which Leibniz seemed already aware and which was certainly present in the mind of Boole (sections 1613 and 162).

223 Mathematical axioms not self-evident

Not only has the conventional and transitory nature of axioms come to be recognized, but the problem of their self-evidence has been shown to be nugatory and one which should not be posed. It is not at all evident, as Reichenbach stresses, that a straight line is the shortest distance between two points, and there is no way, nor even need, to prove it. It is the same with every other case. Even those writers who do not completely reject the help of evidence are agreed, however, that they must free themselves from its "oppressive weight" — "free," adds Scholz, "in the sense that one first of all chooses axioms so that

---

[1] H. Reichenbach, *op. cit.*, pp. 125–6.

"The precedence of synthetic intuition over analytic demonstration in mathematics seems to have continued during the flowering of Greek geometry. In a letter to Eratosthenes, Archimedes thus expresses himself: 'Many things which first became clear to me through mechanics were later demonstrated by geometry, since the treatment by the mechanical method was not yet based on proof; it is in fact easier to find the proof when a representation of the problems under review has first been reached by the mechanical method, than to find it without a preliminary representation'." (From E. Colerus, *Piccola storia della matematica da Pitagora a Hilbert*, Einaudi, Turin, 1939, p. 72.)

they are immediately felt intuitively, and then extracts from them everything that can be extracted by a proper system of rules, instead of becoming involved in insoluble questions about the basis for the feeling of evidence and the criteria to go with it."[1]

Since a logician is concerned with the self-consistency of developments from axioms and not with the evidence in their favour, axioms have been demoted from the rank of principles to that of beginnings, and are made subject to further observations which may disprove them, or to revocation of the conventions which they satisfy.

224   Empirical foundations of scientific axioms

For the moment we cannot discuss the relations between abstract and practical disciplines and between their axioms without getting on to applied mathematics (see sections 18 and 181) or on to the more general idea of logical synthesis, the deepest concept so far worked out to find a relation between the two classes.

But we can add some further remarks. Newton's concept of mass has been referred to above; set out in axiomatic form, it certainly has an abstract character and a quantitative content, but Newton, far from considering it as an instrument to be developed tautologically, was thinking rather of its applications to natural facts.

Still more significant is the relativity axiom of the constant velocity of light. Intuitively, one should admit that velocities can be added and subtracted, and that therefore a man on a long, very fast missile who moved from his place, even slowly, in the direction of motion, would increase his velocity with respect to a stationary observer. But, in relativity theory, this does not happen with regard to the velocity of light. The constant velocity of light is a primary fact accepted by relativity theory, without the possibility of reference to evidence, and it is simply a useful premiss for the consistent development of the logical system of which it is a part.

Bridgman[2] (already mentioned in section 1421), in analysing the axiom of length, has pointed out that the geometrical definition, in relation to a segment of a straight line joining two points, is not enough to free it from the measuring technique on which it depends. "The concept of length", he says, "is therefore fixed when the operations by

---

[1] H. Scholz, *Storia della logica* (1931), Italian edition, Silva, Milan, 1961, p. 132. E. Agazzi, *Introduzione ai problemi della assiomatica*, Vita e Pensiero, Milan, 1961, p. 12.

[2] P. M. Bridgman, *The Logic of Modern Physics*, Macmillan, New York, 1949, p. 5.

which length is measured are fixed; that is, the concept of length involves as much as and nothing more than the set of operations by which
length is determined. In general, we mean by 'concept' nothing more
than a set of operations: *the concept is synonymous with the corresponding
set of operations"* — and, one could add, of corresponding linguistic
formulations.

These and other points lead to the thought that not only are abstract
disciplines constructed and developed tautologically, bringing out the
implications of the original axioms, but the factual sciences also base
their premises on axioms, from which they deduce lines of reasoning
to be tested against reality. This matter will be discussed later. But
it has so far been important in another respect, since it helps to
strengthen the objections to the *a priori* interpretation of mathematical
axioms, showing that there exists an entire gamut of them, ranging
from remote factual premises of axioms which are now difficult to
retrace, to those whose concrete nature is still obvious, although they
are conceptually abstract and almost void of objective content.

23   The idea of number

For a quick appraisal of the conventional nature of mathematical
axioms, let us consider the system of numbers called "natural", whose
time-honoured position clearly reflects their universality.

The following expression:

$$N = x_1 p^{m-1} + x_2 p^{m-2} + \ldots + x_m p^0$$

is suitable for formulating as many numerical systems as one wishes,
where p is the base of the notation, the x's are the digits composing
the number N, beginning with the first on the left, and m their number,
subject to $m \leqslant p$-1. It can immediately be seen that the formula supplies all the numbers of the usual decimal system when $p = 10$.

But it is also true that every number containing figures not above
8, 7, 6, etc., can be interpreted as belonging to a nonal, octal or sep-
tenary system, etc. It is likewise possible to express with the formula
numbers belonging to an order above the decimal, provided that
arbitrarily new symbols are created to denote numbers above 9.[1]

---

[1] In old English there were the remnants of a vigesimal numeration in which
$\frac{xx}{iii}$ was written for 60 (see M. Boldrini, *Statistica, teoria e metodi*, Giuffrè,
Milan, 1962, p. 128). In French this numeration is still used for the numbers
70–99. In a passage from Molière's *L'Avare*, a female character, flattering a
very old lady, says: "Vous passerez les six-vingts". Examples could be
multiplied as confirmation of the purely conventional basis of numeration.

It is usual today, for reasons already mentioned (section 162) to use binary numeration, in which only the numbers 0 and 1 are used. Thus the number 110100010, which has a meaning in the decimal system, corresponds to the decimal number 418 when interpreted in the binary system. Summing in the usual way, one has in fact:

$$(1 \times 2^8) + (1 \times 2^7) + (0 \times 2^6) + (1 \times 2^5) + \ldots = 418.$$

Thus the ordinary notation of numerical calculation in the scale of 10 is seen to be arbitrary in the sense that the usual denary formula $2 + 2 = 4$ becomes, in binary, $10 + 10 = 100$, and in ternary, $2 + 2 = 11$, and so on. No confusion results from this, provided one accepts the fact that the sum of two numbers can be expressed by many possible conventions.

231   Non-Euclidean geometries

The sweeping changes which took place not long ago in the basis of geometry are also relevant in this context, though their epistemological impact is really much wider.

As is well known, the starting-point consists in excluding from the axioms of Euclidean geometry the postulate of parallels, nowadays called the postulate of the right angle, in consideration of the fact that in a plane a perpendicular to a straight line also forms right angles with its parallels. Euclid formed the postulate of parallels in a hypothetical and not an apodictic way, and it has always been held suspect by geometricians. Gauss seems to have been the first to investigate the consequences of rejecting it, but he did not publish his results; and the findings of the Hungarian G. Bolyai remained almost unknown. When in 1826 the Russian Lobacewski, changing the same postulate, constructed the first non-Euclidean geometry, the scientific world greeted it as a revolution.

Consider the surface of a large sphere, which might be the terrestrial globe. Each great circle lies in a plane passing through the centre, and each arc of it, called a "geodesic", measures the least distance along the surface between the end points of the arc. Thus, a geodesic corresponds, on a sphere, to the shortest distance between two points which, in a plane, consists of the segment of a straight line which joins them. Now take two points on the earth's equator and pass through them two meridians, and then at an equal distance from the equator let the two meridians be intersected by a further great circle. The four great circles thus traced will contain a trapezoidal surface enclosed by four geodesic lines forming two right angles at the equator and two obtuse angles at the other corners. This illustrates a feature of one kind of non-Euclidean geometry, known as that of the obtuse angle. Clearly this geometry excludes the parallel postulate while accepting all the other Euclidean postulates and, naturally, differs in many of its deductions from classical geometry.

Some years later, Riemann considered not a sphere, but a surface produced by the rotation of a special plane curve called a tractrix (roughly the shape of two trumpets soldered together by their mouthpieces and extending to infinity in both directions), and he formulated a geometry analogous to that of Lobacewski but in which those figures which, for the sake of clarity, can be called pseudospherical trapeziums, have two internal right angles and two acute angles. Between these two opposite cases, that of the obtuse and that of the acute angle, Euclidean geometry, accepting the postulate of parallels, appears as an intermediate limiting case.

We have said that Euclid was compelled to adopt the parallel postulate, although regarding it as hypothetical. One may surmise that this arose from the mistaken opinion of the Greeks that the earth was flat. If this was its empirical origin, there would be some support for the claim that such a postulate is not as obvious as it seems. Consider, furthermore, that the non-Euclidean geometries, and especially that of the acute angle, have greatly contributed to the mathematical formulation of the relativistic conception of the world, and one will conclude that science has progressed not through the absoluteness of postulates but rather through their changes.

2311   Gauss's criticism of Kantian synthesis

We have seen that Gauss discovered the principles of non-Euclidean geometry before Lobacewski. In a letter written in 1844 to his friend Schumacher[1] he lamented the incompetence of many philosophers, both ancient and modern, from Plato to Hegel, and added: "But even with Kant himself, it is not often much better; in my opinion his distinction between analytic and synthetic propositions is one of those things that either run out in a triviality or are false." Bell, who quotes this extract, comments that Gauss was perhaps at that moment thinking that non-Euclidean geometry "was in itself a sufficient confutation of what Kant was saying about space and geometry".

232   The idea of space

To return to the relationship between geometry and mensuration (section 224), and indirectly between abstract mathematics and the

---

[1] E. T. Bell, *Men of Mathematics*, Penguin Books, vol. 1, p. 263.

mathematics of the physical sciences (section 181), we may note that Gauss, after the publication of the work by Lobacewski (about which he was generous-minded enough to be pleased, though it deprived him of a priority which was his due), tried to verify the new geometry by measuring the sum of the internal angles of a geodesic triangle. He found that it conformed to Euclidean geometry.

This procedure is justifiable, provided one accepts some definitive assumptions and, in particular, provided one agrees that two measurements taken in different places and in different circumstances with the same instrument are to be defined *a posteriori* as equal if they are observed to be so. Hence the concept of the relativity of measurement with respect to the postulated invariance of the instrument. But its epistemological importance lies in Gauss's insistence on linking the abstract space of geometry with the actual space of physics. The problem has since been thoroughly studied on the basis of relativity theory, and the conclusion reached is that, with the same assumptions, extending the measurements to cosmic dimensions and taking into account the influence of gravity, space does not conform to Euclidean geometry.

The opinion accepted at the present time (note the limitation) is that physical space and geometrical space do not necessarily correspond. Geometrical space is neither unique nor necessary, but there are as many spaces as there are conceivable geometries; physical space is an experience of the world and, even in its objectivity, varies according to the mental picture in which it is placed and the linguistic propositions in which it is expressed.

Although this conclusion differs notably from Reichenbach's view, it is as well to note his statement that, while the mathematician does not concern himself with the truth of axioms but with their implications in respect to the theorems derived from them, the physicist essentially requires the axioms to be such that it is possible to derive from them a geometrical system suitable for describing the natural world. Mathematical geometry has developed from axioms formulated *a priori*; physical geometry has, however, an analytical character, though it derives its premises from the synthesis of experimental facts. Hence the conclusion—to which we shall return—that "the evolution of geometry culminates in the disintegration of the synthetic *a priori*".[1]

233 The idea of time

Although this may seem a peripheral topic, a word may be added here about the concept of time, whose connections with mathematics

---

[1] H. Reichenbach, *op. cit.*, p. 140.

and the physical sciences are universally known. The principal problem of time is its definition, which is identified with that of its measurement. This can have an analytic character, as when it is introduced as a parameter in mathematical formulae, but it takes on a synthetic character when it concerns the physical sciences. The mathematical axioms of time are developed according to certain purely deductive disciplines such as rational mechanics or stochastic probability. On the other hand, astronomical measurements of time, which concern the natural sciences, always derive from a kind of prior conventional assumption. In fact, whatever may be the physical criterion of measurement, it is subject to considerable objections. Such is the case of the pendulum, whose oscillations, considered as isochronous, serve as a regulator for clocks; such again is the periodicity of the zenith transit time of a fixed star. It is impossible to check whether the isochronism of the pendulum "truly" (not only "theoretically") varies with place and time, or whether, as would be held certain in theory, the sidereal rhythm is subject to secular variations. The expedient adopted to get out of the difficulty consists of "defining" as strictly rhythmical and constant the interval between the successive zenith transits of the same star — "sidereal time" — and comparing the durations of all physical phenomena with this chosen unit of measurement. This means that the reference unit must be axiomatically defined in order to be able to measure duration relatively (section 112).

Other qualities of time, such as simultaneity and succession, which have assumed so much importance in modern developments of theoretical physics, could be equally important points for consideration.

Once again we fall into dualism: on this occasion between mathematical time, whose measures and properties are deducible from non-necessary but unexceptionable *a priori* assumptions, and experimental time, which can be measured by the convention of adopting as the axiomatic units of measure certain natural parameters whose constancy is unverifiable.

234 Whether mathematical axioms are exclusively analytic

It is held by some that, although a mathematical axiom has the conventional nature already asserted, its potential content cannot be exclusively analytical because, if it were so, all propoundable theorems would also be demonstrable.

This does not refer to problems which are not resolvable with ruler and compass (such as trisecting an angle or squaring the circle), but rather to those which are not excluded on theoretical grounds, as might be the case with Fermat's last theorem, mentioned in section 152. But the answer to this point is that one is dealing either with questions badly put or with solutions not yet found. (In Fermat's case one might add — found and then lost again.)

Other very different objections stem from theorems which are certainly true but impossible to demonstrate formally (see that of Russell in section 152)[1]. To understand this, one must go back to a famous paper published some forty years ago by Gödel[2].

24 Gödel's numbers

In section 144 we noted that every statement about mathematical formulae is described as "metamathematics", and that "Gödel's elementary numbers" are the numbers from 1 to 10 which, by convention, he attributed to the ten most usual signs of symbolic logic (section 163). Consequently, anyone who wrote " $\subset$ " instead of the number 3, and ")" instead of the number 9, would be using metamathematics, and the signs would be "numerals" indicating corresponding numbers.

Gödel completed his basic reference table, making the numbers 11, 13, 17, etc. (that is, the prime numbers immediately after 10) correspond to the variables x, y and z (for which numbers or numerical expressions may be substituted), and the squares and cubes of these numbers correspond respectively to the variables called "propositional", namely p, q and r (for which sentences may be substituted)

---

[1] Russell's theorem reminds one of a hypothetical statement made by Archimedes in his dissertation on the infinite: some maintain "that there has never been spoken a number so great as to exceed the number of the grains of sand". This extract, with which the demonstration of the infinitely great begins, is quoted by E. Colerus, *Piccola storia della matematica da Pitagora a Hilbert*, p. 67.

[2] The celebrated memoirs of K. Gödel, *Über formal unentscheidbare Sätze der Principia Mathematica und verwandter Systeme*, 1931; translated into Italian and published as an appendix to the volume by E. Agazzi, *Introduzione ai problemi della assiomatica*, p. 203. Agazzi analyses thoroughly the so-called Theorem of Gödel, its extensions, and the lessons to be learnt from the need of consistency and completeness in an axiomatic system. See also E. Nagel and J. R. Newman, *Gödel's Proof*, New York University Press, 1958; H. Jeffreys, *Scientific Inference*, Cambridge University Press, 2nd edn, 1957, pp. 9 and 18; W. van Orman Quine, *op. cit.*, p. 297; L. Geymonat, *op. cit.*, p. 83; and Francesca Rivetti Barbò, "L'antinomia del mentitore nel pensiero contemporaneo da Peirce a Tarski", *Vita e Pensiero*, Milan, 1961, pp. 288 *et seq*. In addition to Gödel's work on restrictions in logical systems, these various authors deal with those of A. Church and the objections to the consistency of arithmetic by G. Gentzen.

and to the predicatives P, Q and R (for which predicates may be substituted).

By adopting Gödel's numbers 1, 2, . . ., 10, 11, 13, 17, 11^2, 13^2, 17^2, 11^3, 13^3, 17^3, as exponents of successive prime numbers from 2 onwards, every formula of symbolic logic (written so as to contain only the numerals corresponding to the preceding elementary numbers) can be reduced to the number obtained by multiplying the aforesaid powers by each other. The result is called a "Gödel number". Thus, on the basis of the table in section 163, Gödel's elementary numbers appearing below can be made to correspond to the symbols of the logical formula of Peano's second principle (cf. sections 15 and 163),

$$(\quad \exists \quad x \quad) \quad (\quad x \quad = \quad s \quad y \quad)$$
$$8 \quad 4 \quad 11 \quad 9 \quad 8 \quad 11 \quad 5 \quad 7 \quad 13 \quad 9 \tag{1}$$

Working as explained above, we obtain the following "Gödel number":

$$n = 2^8 \times 3^4 \times 5^{11} \times 7^9 \times 11^8 \times 13^{11} \times 17^5 \times 19^7 \times 23^{13} \times 29^9. \tag{2}$$

Similarly, Gödel's elementary numbers shown below correspond to the proposition "Socrates in mortal", which, as seen earlier, can be translated into metalinguistic symbols:

$$(\quad \exists \quad x \quad) \quad P \quad (\quad x \quad) \tag{3}$$
$$8 \quad 4 \quad 11 \quad 9 \quad 11^9 \quad 8 \quad 11 \quad 9$$

and form the Gödel number

$$n' = 2^8 \times 3^4 \times 5^{11} \times 7^9 \times 11^{11^3} \times 13^8 \times 17^{11} \times 19^9. \tag{4}$$

Clearly it suffices to resolve any Gödel number — e.g. the numbers n and n' — into prime factors to reconstruct unambiguously the metamathematical formula from which it is derived: in this particular case, to go from (2) to (1) and from (4) to (3).

241 Gödel's theorem

Gödel made a thorough study of the arithmetical properties of numbers of this type (usually indicated by G), and showed that there are true metamathematical formulae of which one cannot give an arithmetical demonstration. This means that, although G is true, it is nevertheless impossible to decide, on the basis of arithmetic axioms, whether G or its negation $\sim G$ is true.

Consider a formula such as (2) which embodies the symbol y, whose Gödel number is 13. If we replace y by the Gödel number of (2), say n, we have another member of the system, i.e. a new Gödel number.

c

We denote this by sub $(n, 13, n)$. This is a general statement covering any case where we substitute for 13 the Gödel number of the formula concerned. Consider now the expression

$$x \sim \text{dem} \, [x, \text{sub} \, (y, 13, y)]. \tag{5}$$

This means that for all x (being a Gödel number) there is no demonstration of the formula with Gödel number y, i.e. that the formula with number y is undemonstrable. Suppose the Gödel number of (5) is n and consider G, say,

$$x \sim \text{dem} \, [x, \text{sub} \, (n, 13, n)]. \tag{6}$$

This is obtained from (5) by replacing y with its own Gödel number and hence has Gödel number sub $(n, 13, n)$. Thus the statement with this number asserts its own undemonstrability.

Hence, if G were demonstrable, $\sim G$ would also be so.

Thus the axioms of arithmetic must be judged inadequate inasmuch as they do not allow of a decision being made between the affirmation of G and its negation, $\sim G$. Even if the axioms are compatible, some questions are undecidable in the sense that they cannot be formally deduced from the axioms. It would not be enough to add Gödel's number G as an integrating axiom, because the new system thus obtained could be shown to be still insufficient to supply all the truths of arithmetic.

On a note of resignation we may say that "the vast region of arithmetic truth cannot be reduced to a systematic order which sets forth, once and for all, a collection of axioms from which can be formally deduced all true arithmetic propositions".

242 Fertility of Gödel's analysis

As described above, Gödel's results are disconcerting, but this account of them, which would once have been unacceptable and perhaps full of serious consequences, the modern mind can take in its stride, because scientific and mathematical doctrines are enriched, rather than impoverished, by the increase and progress of knowledge. This indeed is the opinion of Quine and Geymonat; the latter, generalizing further, says that the limitations of arithmetic revealed by Gödel and also by Church "do not, in fact, constitute a check to human rationality; rather can it be said that the strength of such rationality shows up well by having discovered, through logical reflection, such peculiar limitations, beyond the grasp of ordinary intuition. It is the same story in algebra, where it was shown to be impossible to solve explicitly polynomial equations of degree higher than four; this result, apparently negative, has been surprisingly productive and has increased our range of algebraic operations, opening the way to new and higher problems."[1]

Similar circumstances to those recorded by Geymonat occur throughout the history of mathematics. In the early days of the infinitesimal calculus it was held that continuous functions are differentiable, and Weierstrass caused much ado when he showed that numerous continuous functions have no tangent and therefore cannot be differentiated. But from such discoveries, and specially through the work of Weierstrass, Riemann, Hamilton and others, the general theory of functions and complex variables took shape.

Two general conclusions can be drawn from the above: the first, relative to the subject of this chapter, is that every shadow thrown on to the firmament of mathematics confirms that its basis is terrestrial — that is, human and not astral; the second, more general, is that in keeping with contemporary scientific criticism (already examined in Chapter 1) no true discovery, even though it goes against accepted opinion, can constitute a backward step in the enrichment of knowledge.

25 The transient nature of axioms

All these considerations lead to the down-grading of axioms of the abstract disciplines from *a priori* truth; but the implications of this result would probably have remained latent without the renewed impetus which was given to scientific thought in the first decades of the twentieth century. In Chapter 1, in particular, we considered the radical change of ideas introduced by Einstein in his critique of the concepts of space and time, and by Planck with the quantum interpretation of the phenomena of emission and absorption of radiant energy by a black body[2]. This breakthrough caused a crisis for the dogmatism of all previous science and substituted the idea that scientific theories are not objective or detached observations concerning the world, nor discoveries of its mathematical system, but simply linguistic expressions dictated by convenience rather than necessity.

---

[1] L. Geymonat, *op. cit.*, p. 84.

[2] A simple account of the advance made by Planck on Kirchhoff's ideas about a black body can be found in R. Girelli, "Max Planck", in *La Scuola in Azione*, 1959–60, No. 13, which also contains extracts from his autobiography and from other works of his. Useful reading is the biography edited by E. Camatini, "Albert Einstein", *ibid.*, No. 9, accompanied by anthological extracts.

251   An illustration from economics

Epistemologists usually select their examples from the physical sciences, clinging to earlier ideas on the regularity of the phenomena studied in these sciences. Lord Kelvin's approach (discussed in section 111) crops up in a variety of ways, now less insistently but always as a limitation. Indeed, as regards axioms, nothing is more instructive than the social sciences, either because their growth is recent enough to permit of reference to their origins, or because of a greater willingness in these sciences to reject and adapt, and to accept criticism. Every academic treatise on economics opens with very general premises (expressed as axioms or, what amounts to the same, as definitions) concerning demand, supply, preferences, etc. The statements are undoubtedly experimental in origin, but their conceptualization and consequences are developed in a completely abstract form in so-called static economics, in which the theorems of utility, of demand, of indifference curves and so on can never be verified in practice. This is because economics leaves out of account the time dimension and the accidental circumstances which influence economic decisions[1]. With the increase of practical knowledge, and under the impetus of research into trade fluctuations and economic development, interest has been focused on the dynamic aspect of economics. The modern attitude has shifted from considering the view and actions of a single individual to the broader global outlook, and a social or "macro-economics" has emerged as of greater importance, though without abandoning the economics of the individual, or "micro-economics". The axioms from which macro-economics start are clearly quite different. Here the pivotal points are the *a priori* concepts of production, revenue, consumption, savings, and investment; and from these ideas emerged a

---

[1] See for example, the *Principi di economia pura*, by Maffeo Pantaleoni, Fratelli Treves, Milan, 1931 (first published by Casa Barbera, Florence, in 1889). The first chapter begins as follows: "Economic science consists of the laws of wealth systematically deduced from the hypothesis that men are moved to act solely by the desire to attain the greatest possible satisfaction of their needs with the least possible individual sacrifice." The second chapter continues: "The economists' hypothesis by which men . . . would be moved solely by the desire to attain that greatest possible satisfaction of their needs . . . through the least individual sacrifice can be postulated as a fact, and it is irrelevant to expect this fact to correspond to reality." More recent economists start from analogous premises: see C. Bresciani-Turroni, *Corso di economia politica*, vol. 1; *Teoria generale dei fatti economici*, 3rd edn, Giuffrè, Milan, 1956.

whole new series of problems, envisaged in a way that had escaped the classicists[1].

One would like to continue this excursus, but it would be a digression. Suffice it to say that economists held firm to the postulates and classical developments until about the time of the great depression after World War I; but with the dynamism which came to enliven all the sciences between the two wars and later, and with better insight into the nature of phenomena and a more open mind about subsequent developments, they steadily proceeded to re-appraise and revolutionize all standpoints — factual, interpretative, and linguistic (section 5313).

252 Axioms in a system due to Dingle

Apart from Poincaré and Mach, perhaps nobody in recent years has brought into clearer evidence than has Dingle (already mentioned in section 171) the connection, in science, between observable data and conceptual elaboration, which latter includes axiomatic premises[2]. After criticizing the view of science as a progression of concepts slowly extended and integrated, he maintains that, on the contrary, experimental knowledge consists of disjointed elements which the mind puts together so as to produce comprehensible descriptions. The simile he

---

[1] The trend towards macro-economics in the science of modern economics is attributed to the so-called "Keynesian revolution"; see J. M. Keynes, *The General Theory of Employment, Interest and Money*. For a further appreciation of the Keynesian theory and its developments, see J. A. Schumpeter, *Ten Great Economists from Marx to Keynes*, Oxford University Press, 1951; and, by the same author, *History of Economic Analysis*, Allen and Unwin, London, 1955, especially Chapter 5 of Part V. Also F. di Fenizio, *Studi keynesiani*, Milan, ed. L'*Industria*, no date; *Studi keynesiani*, Istituto di Economia e Finanza della Facoltà giuridica di Roma, Giuffrè, Milan, 1953; L. R. Klein, *The Keynesian Revolution*, Macmillan, London, 1956.

It is true, however, that the bridge between the axiomatics of macro-economics and that of micro-economics is not necessarily broken; part of the latter is implicit and tacitly accepted, but its principles are no longer brought forward in evidence, nor do they serve as starting-points for further deductive developments: they remain a remote premiss and have given way to a new set of axioms which have now come into the limelight.

For developments in economic science, some of its main problems and its logical construction, see also Tjalling C. Koopmans, *Three Essays on the State of Economic Science*, McGraw-Hill, New York, 1957; K. Kubihara ed., *Post-Keynesian Economics*, Rutgers University Press, 1954; K. Kubihara, *Introduction to Keynesian Dynamics*, Columbia University Press, 1956.

[2] H. Dingle, "The rational and empirical elements in physics", in *Philosophy*, London, vol. 13, April 1938.

repeatedly uses is to compare scientific construction with a jigsaw puzzle.

The research worker looks at the jumbled pieces of the puzzle, which in themselves are neither true nor self-evident; the mind rationalizes them and, as will be shown later, introduces subjective connections (cf. sections 2521 and 364). With previous knowledge and a specific aim, it composes them into a coherent pattern. In contrast to the child's game, there is no known over-all design which suggests one result and one only; on the contrary, each group of pieces and each concept embodied in that group depends on the greater or lesser consistency achieved. Thus it is conceivable that the same elements may be used to symbolize, either singly or together, two or more different pictures, such as an electron which is interpreted sometimes as a wave, sometimes as a corpuscle (section 1421). These interpretations are more or less true, depending on their context.

From the viewpoint which interests us at the moment, it should be added that the solver of a puzzle would be deceiving himself if he thought he could commence his work without some preconceived idea. There will be a few pieces which will allow him to make a start, and then the design will develop piece by piece. The initial idea — whether rational or not — will constitute the "axiom" of the worker, the element determining the whole deductive procedure. And since the processes of observation, research and thought develop together, facts stimulate ideas, and the mind seeks to fill in the missing links so as to make the pieces of the design more or less intelligible. (Everybody knows how the periodic table of chemical elements was left by Mendeleyef with gaps in it which were gradually filled in; and of the discoveries of stars and elementary particles already foreseen through the calculations of astronomers and physicists.) These things are established and communicated through language.

2521   Corollaries of Dingle's system

Dingle's scheme merits further elaboration. In fact, the pieces of the puzzle never fit perfectly. They may sometimes be elementary sensations, in Mach's sense; at other times partially worked-out combinations, if for no other reason than the inequality of all natural facts. ("A rose" does not exist that is equal to another, nor "a sunset" of the colour of today's, nor "a loved woman" who is personified in Laura or Fornarina or Mathilde Wesendonk.) Hence, to fit these elements together, the scientist must mould them — following certain rules — to his mental picture, and submit them to the necessary experimental tests.

As the professional mosaic-worker divides up his design into squares, assembling the colours in a relatively small number of shades, grouping the pieces according to colour and only then sets to work, so the scientist assembles, moulds and plans the data of experience and translates them into intelligible linguistic terms. Although Dingle doubts this (section 364), the procedure is in keeping with the whole actual trend of science, as Wittgenstein already knew (section 5643). Phenomena based on the same elements can be considered from different viewpoints which may lead to the solving of further problems. To continue with the comparison, it is possible to obtain the most varied figures with the same pieces of mosaic cut in geometric shapes: for example, with six equilateral triangles of the same size it is possible to construct a hexagon, a star, a parallelogram, and still other designs.

The criterion of truth is sometimes based, together with logical consistency, on limits fixed *a priori*, or on the achievement of practical ends, or on agreement with evidence or previous statements, or on some other reason. From the four letters AOMR, 24 permutations can be made; some, such as AMOR, ORMA, RAMO, ROMA, MORA, can be "true" as compared with "false" ones, because the former but not the latter correspond to Italian words; but other arrangements could be "true" or "false" according to whether they correspond to the winning tickets in a lottery. Similarly, it was stated earlier that propositions can be judged sensible or otherwise according to the linguistic convention of the context (cf. sections 143 and 1431).

253 An illustration from chemistry

Consider again the periodic table of chemical elements. Mendeleyef actually constructed it as one puts pieces together. The chemical elements (considerably more numerous today than those known in 1869) are written in the rows and columns of a table, from hydrogen, H, the lightest, to uranium, U, the heaviest. The columns form "groups" of elements with some similar properties, and the rows form "periods".

The first row contains 9 elements, but since H remains apart, the period is in reality formed from eight elements only, from the second to the ninth, that is, from helium, He, to fluorine, F. The following eight elements, up to number 17, form another period in order of increasing atomic weight, from neon, Ne, to chlorine, Cl. One by one they repeat certain properties of the corresponding elements of the previous period, ending, as this latter one does, with another halogen.

In the third period the rule is modified by adding another seven elements to the first eight elements which follow and which end with manganese, Mn. Thus is obtained a first "long period", which ends with bromine, Br, that is, the third halogen. Then follows a second long period, in which the eighth place is taken by technetium, Tc (corresponding to the Mn group), and after another seven elements we come to iodine, I, the fourth halogen. And so on, but always with more complex adaptation. The properties of elements in the groups correspond only approximately.

Mendeleyef's table should be considered, however, as a combination of fact and reasoning, and is justified on grounds of utility rather than consistency.

It is difficult to imagine the axiomatic starting-point from which Mendeleyef constructed his periods; clear or obscure, recent or remote, one must suppose that he had a starting-point if the sequence is to be anything more than a simple description.

254   Conclusions on the conventionality of axioms

The conception here set forth of science in general and of axioms in particular is part of those essentially modern revolutionary achievements which were mentioned in Chapter 1. It accentuates the role of the interpreter in science, with his accompanying intellectual and logical attributes. Consequently it is opposed to that passive objectivity which the positivistic science of the nineteenth century mistakenly thought it had achieved, and to the epistemology and metaphysics built upon this.

The advent of the interpreter also led to the rejection of the idealistic attitude which, reducing everything knowable to the subjective level, excludes the useful interaction between the knower and the thing known, by means of which science laboriously builds its edifice, controlling and renewing its criteria of truth. This being so, the ideas about mathematical axioms gain greater prominence. It is no longer possible to assign them an *a priori* validity, given their conventional formulation, whether this is reached by intuition (as Hilbert thought) or from more or less remote factual sources. Conversely, the axiomatic principles of science lack the evidence and the detached objectivity which gave the scientists of the past such satisfaction, and on which they built philosophic doctrines that have not stood up to criticism. For instance, the constancy of the relativistic velocity of light is by no means obvious, but common sense accepts it all the more because, given its exceptional magnitude, the velocity is not influenced for practical purposes by small additive quantities. (By comparison with the velocity of light, that of sound and of missiles is trivially small — section 224.) But this practical justification takes away rather than adds evidence to the relativistic axiom.

26 Axioms and axiomatic methods

To round off our survey, it should be pointed out that it is one thing to speak of axioms in general and another of the axiomatization of mathematics.

Now that the conventional character of axioms has been made clear and the rules of transformation in the abstract subjects of logic and mathematics are fixed, there is a tendency in demonstrating theorems, to exclude the introduction, overt or covert, of new extraneous elements, on the pretext that they would be "obvious", or "evident", or what have you. This is known as the "axiomatic method". "To axiomatize a theory", writes Geymonat[1], "means to establish that every concept, property and operation will be used, in a given context, only in the precise sense in which they were implicitly defined by the axioms of the theory. These axioms take on the function of general syntactical rules for that particular scientific discourse."

It might at first sight seem possible to axiomatize the concrete sciences in the same kind of way, as though they had a strictly deductive nature. This point will be better appreciated later. But on the basis of common experience, one can understand that there are factual disciplines which, at least in part, seem to be completely detached from any functional contact with experience. Such is the case with mathematical economics, whose purely theoretical apparatus consists of more or less complex theorems and developments of that kind of applied mathematics mentioned in Chapter 1. Theoretical physics is in a similar position; it exists and develops deductively, with apparent detachment from the facts and laboratories of experimental physics. Other disciplines sometimes put forward particular problems in a purely abstract form and attempt to solve them in a mathematical or logical way. Certain aspects of demography[2] are examples, and also that very modern branch of economics known as econometrics.

It can be shown that these ideas also apply to other branches of scientific knowledge. But not for this reason should axiomatization be selected as the ideal road to progress. Axiomatization is the progressive conquest of abstract disciplines, but, unlike these, the factual sciences base their rigour on rather different methods — to be considered later — the proof of whose validity is written into the story of present-day scientific and technical progress.

---

[1] L. Geymonat, *op. cit.*, p. 70.
[2] M. Boldrini, *Demografia*, Giuffrè, Milan, 1956, p. 434.

C*

261   Science and neo-humanism

From the foregoing it emerges that modern science, both abstract and concrete, has made a decided return to a form of subjectivism. This is certainly different from the kind the rationalists or the idealists thought of, but it is above all in antithesis both to the objectivism of the empirical philosophers and to positivism. Because of the emergence of the individual at the centre of study, it is likewise in antithesis to the solemn and detached dignity of the Kantian critique. Convergence of the various factors we have been examining — from the questioning of the basis of Euclidean geometry to the relativistic revolution — has led to the modern position, which speaks of neo-humanism as of a characteristic peculiar to current scientific thought[1]. Among the many documents which could be cited in support of this concept, it is enough for the moment to select one of acknowledged authority. "The formulation of a problem", wrote Einstein[2], "is often more important than its solution, which is sometimes reduced to a mere question of mathematical or experimental ability. To propose new problems, to point out new possibilities, to consider old problems in a new light is what truly requires creative imagination and marks real scientific progress." They are illuminating words, even though they reflect an outlook which has been superseded, as will be seen later (section 314). Einstein, unlike others, never willingly conceded that imagination itself could create a scientific reality which was capable of producing results, although not objectively verifiable in the old sense of things that could be touched and seen.

In spite of this, Einstein's ideas, which he reaffirmed many times, led to new and splendid achievements and added a new dimension to our knowledge of Nature.

27   Conclusion

To return finally to Kant, who in all modern epistemological works seems to be the Great Defendant: one should not only concede him the merit of having freed studies from the obstructive dialectic between rationalism and empiricism, but also, with logical *a priori* synthesis, to have sown such fertile seed that not even the numerous critics of the

---

[1] These concepts are fully developed in the lecture, inaugurating the course on Statistics, given in Rome University on May 3rd, 1956: M. Boldrini, *Neo-umanesimo e statistica*, mimeographed edition.

[2] A. Einstein, *Conceptions scientifiques, morales et sociales*, French edition, Flammarion, Paris, 1952, p. 106.

last fifty years have succeeded in smothering. It is true that Gauss already had doubts as to the acceptability of the Kantian critique (section 2311), and in the opinion of von Mises and Reichenbach it should have been overthrown by now, "since, according to Kant and the idealist philosophers, there are without doubt fixed concepts, independent of our will, which have existed before all linguistic conventions, and whose delimitations can be studied more or less precisely".[1] This assertion contrasts with the conventions of today, universally accepted both in abstract disciplines and in those concerned with natural science (section 3326).

We merely re-emphaize that in the structure of modern science there is a human dimension. There is a factual element in mathematical axioms, and a subjective content in scientific axioms. But the Hilbertian hypothesis of an intuitive basis in the first, and probably in the second, would suggest that great mathematical ability plays a part in determining the origin of axioms. This ability is part of the mysterious working of the mind and of the more general functioning of the human organism. There are human beings, not easy to define, whose psychological and biological makeup is so constituted as to place them outside accepted conventions. They are, and are not, Nature; science makes use of them but does not know them. It is as well, therefore, that the general statements of mathematics and science do not coincide with the Kantian *a priori*. But they cannot free themselves with certainty from that mysterious something which is always immanent in the external world, in the inwardness of a human being, and in the relations between him and the world.

The conclusion — if we have really reached that stage — is this. Once science ceased to be thought of as absolute and was related to mental concepts based on observation, a great step toward clarification was achieved, bringing with it bold and unlimited progress. But the name of Kant still rises high in the firmament of science, not only for the influence that he exercised at the beginning of the nineteenth century and for his part in eliminating the dualism between *res cogitans* and *res extensa*, but also for many other stimulating ideas, however attenuated and misinterpreted they may have become in the crucible of contemporary scientific thought.

---

[1] R. von Mises, *op. cit.*, p. 167.

3

Deduction and induction

31 Wittgenstein and the logical concept of the world

When Wittgenstein wrote down his basic axioms, referred to in sections 141 and 151, he formulated in modern terms the outcome of a revolution in thought which had been developing from Descartes to Leibniz, Hume, Kant, Mach and Russell. As with all axioms, those of Wittgenstein lack *a priori* validity and evidence; they are, on the contrary, a decisive point of departure for a development which recent history has shown to be exceptionally dynamic and penetrating.

These axioms maintain that the world is simply a (dissociated) whole of elementary events, ordered and built up by thought and expressed in words. There is no *systema naturae*, no cause and effect; no objective laws to discover, but simply an immense field of work for observation and logic: a new reality which has not yet revealed all the explosive potential concealed in it.

311 Mach's unification of objective and subjective

At a time when the moderns had begun to reject, in the name of science (but not in that of metaphysics), the opposition of the two worlds of ideas and of facts (section 172), the name of Ernst Mach came to the fore.

For Mach, "the question whether the world is real or only a dream of ours has no scientific value".[1] Knowledge of the permanent elements

---

[1] E. Mach, *Analisi delle sensazioni*, Italian edition, Bocca, Turin, 1903, p. 13. This celebrated work was first published in 1886, and was followed by many other editions, the last being in 1906. The Italian edition, dated 1902, carried a preface by the author which opens with this outline: "The idea continues to gain ground that science should confine itself solely to the clear and succinct description of positive facts. There follows logically from this the elimination of all the various hypotheses which are outside the control of experiment, especially metaphysical ones. From this point of view, in

of bodies reaches the ego "through sensations". In this way the former and the latter are integrated, so that instead of using the expressions "elements" or "complexes of elements", the parallel terms "sensations" or "complexes of sensations" can be adopted[1]. Thus is created the link between the physical and psychological, between object and subject. And since every body is a complex of elements, it is likewise a complex of sensations[2].

Mach writes as follows: "If we break up the whole material world into elements which are, at the same time, elements of the psychological world and, as such, are called *sensations;* if, finally, we admit that the special task of science is to seek out the connections, relations and mutual dependence of these elements of like nature in all fields, we can expect on good grounds to give this representation a unique monistic basis, and to remove the useless and mistaken dualism. By considering matter as something absolutely persistent and immutable, the connection between the physical and psychological breaks down."[3] And again: "We fully understand a natural phenomenon . . . when our thoughts exhibit to us the total of the relative sensory facts, . . . presenting them as known, so that we are not overwhelmed by them."[4] "All subsidiary representations, laws, and formulae are only the quantitative measure of my sensory representation. The latter is the aim, the former are the means."[5]

The whole of Mach's theory turns on these basic ideas, from which the author draws a copious series of deductions.

To begin with, there is a complete reversal of the notions of physical reality. Every physical phenomenon is, in a certain sense, a subjective–objective reality, an "adaptation of thought to investigation", "an

---

the widest field embracing both physical and psychological phenomena, one reaches the concept that the so-called sensations form the integral parts, the elements of all possible phenomena. In consequence, phenomena consist of different ways of combining these elements in their reciprocal dependence. A whole series of illusory problems is thus removed. . . ." In the same train of thought as Mach, see K. Pearson, *The Grammar of Science* (1892, 1900), Dent, London, 1937, p. 13. For Pearson also, the aim of science is not indeed to explain, but to describe in conceptual terms our perceptive experience. In the development of this thesis he refers explicitly to Mach. Pearson's *Grammar* has had a great influence in England, partly as a result of his fame as a statistician.

[1] *Op. cit.*, p. 19.
[2] *Op. cit.*, p. 380.
[3] *Op. cit.*, p. 360.
[4] *Op. cit.*, p. 363.
[5] *Op. cit.*, p. 362.

activity of reactions, which enriches a fact with a new sensory element".
In this framework, what will matter become? Simply a symbol of
thought. This is always true, and "it must be much more valid for the
hypothetical and artificial atoms and molecules of physics or chemistry.
Nevertheless they still have a value in their own special domain of
application. They remain economical symbols of physico-chemical
experience, but as with algebraic symbols, we cannot hope to get more
from them than they contain."[1]

Evidently Mach does not face the main question of establishing
what is abstract and what is concrete (nor the corresponding linguistic
expressions — sections 133 and 15) and hence avoids the problems of
rationalism and objectivism, of the tautological sciences and the
factual ones, of the formal modes of expression and those of content
(section 172), and of the complicated interweaving of thought which
has been built up on this theme from Descartes' *cogito* onwards. He
confines himself to the problems of the physical world and indicates
the path to be taken by modern scientific thought, in which object and
subject are integrated to form a single, inseparable reality.

3111   Mach's criticism of the concept of cause

The importance which the subjective element assumes in Mach's
philosophic system is likewise at the root of Hume's earlier rejection
of the concept of cause. "The opinion (says Mach) held in olden days on
causality is something of a muddle: to a given quantity of cause corresponds a quantity of effect. This opinion depicts a world as primitive, as
unsophisticated as the theory of the four elements. This appears indeed
from the word 'cause'. Connections in Nature are not generally so
simple that, in a given case, there is a cause and an effect. For several
years now I have been trying to substitute for the idea of cause the
mathematical concept of function: that is, *interdependence of phenomena*,
or more precisely *interdependence of the distinctive characteristics of
phenomena*."[2]

And a little further on, insisting on the necessity of substituting the
functional relation for the causal relation, he adds polemically: "I
imagine to myself what every modern researcher into Nature must
think who bears in mind (for instance) the concepts of Mill on the
method of experimental research: he would not get as far as attempting
to apply them, even in the most provisional example."[3]

---

[1] *Op. cit.*, p. 359.
[2] *Op. cit.*, p. 110.
[3] *Op. cit.*, p. 114.

3112   Mach's criticism of the concept of matter

The effect of Mach's ideas on the problems of science is still evident today. He affirms that the common concept of matter can remain "for ordinary and practical use," together with the concept of measure, but that instead of metaphysical concepts, Nature admits only empirical ones. "Yet Science suffers no detriment if, in place of a *quid* (matter), rigid, sterile, persistent, unknown, we postulate, by means of physico-physiological research, a persistent law."[1] Clearly, this conclusion could not now be accepted other than in the most general way, that is, not as a new dogmatism but simply as the affirmation of a subjective–objective thing-known which can be contrasted with the knowable asserted by positivistic science. This being so, Mach's thought remains as one of the definite bases of current concepts.

3113   Later opinions on Mach's thought

At the end of the nineteenth century Mach's ideas provoked obvious reactions, but the same motives for which they were then contested are those which today make them acceptable. Obviously, also, von Mises, an adherent of Viennese neo-positivism, insists on the close connection between the "elements" of Mach and protocol propositions (section 172). There is, again, an echo of Mach's thought in axioms no. 1, 2, 3 and 4 of Wittgenstein (section 141).

Scientific agreements and disagreements over Mach's work have nothing to do with the repeated praise heaped upon him by Croce, the Italian philosopher. The object of this was to validate the individual–universal category in history and logic, as opposed to the pseudo-category of the real–general in science. "The physical sciences", wrote Croce, summarizing Mach, "no less than zoology and botany, have as their sole principle the description of natural facts, in which there are never equal cases. Equal cases are moulded from the schematic imitation of reality; and this is also the origin of the mutual dependence among the attributes of facts. To this the principle of causality is reduced, and to avoid foolish fancies it would be better to substitute for it the concept of function."[2]

Croce expresses Mach's thought accurately, although there was a fundamental disagreement between the two authors, because, while the German, with his criticisms of the old epistemology, aimed at setting up a more coherent one, the Italian used them as a defence for his

---

[1] *Op. cit.*, p. 383.
[2] B. Croce, *Logica come scienza del concetto puro*, Laterza, Bari, 1909, p. 385.

almost incredible short-sightedness towards the rapidly developing natural sciences.

312   Intuitive ideas about scientific procedures

In modern science the simultaneous presence and reciprocal conditioning of objective and subjective factors makes for a dynamic equilibrium between them. To come to the core of the problems arising from the position summed up in section 31, remember that the experimental and rational procedures of science run parallel to and accord with the rules and good sense of daily life. It is only because "we know" that night "always" follows day that humanity habitually alternates the hours of work and the hours of rest.

Although death — like the rising of the sun — is considered inescapable, nevertheless in the West survival is almost certain from the age of 20 to 30. Similarly, all reasonable men know that while one can be lucky, one should not count on winning a football pool, or, although each generation is increasing in height, assume that grown men will soon be well over six feet tall. Starting from these assumptions, young men of twenty put their names down for the university and plan their careers; they learn a calling without expecting to win the pools, and architects continue to make doors two metres high, since they are sure that, giants excepted, people will not hit their heads on the architrave.

But to the critic who might ask whence so many certainties spring, it would have to be conceded that nothing which is anticipated is really "known" but only "maintained". It follows that this same inescapability of death cannot be transferred unhesitatingly from the past into the future. Therefore the belief which spread among the disciples through a mistaken interpretation of some words spoken by Jesus, that John the Evangelist would not die (John, 21, 22-23), was not entirely senseless — apart, that is, from the superhuman powers of Our Lord. In the same way, the query already formulated by Hume and repeated by Wittgenstein (6.36311) as to whether the sun might not rise tomorrow, is logically valid (section 151).

313   The notions of hypothesis, axiom, and scientific law

In all the preceding examples it is easy to recognize the existence of a general proposition (the sun rises every day, all men are mortal, etc.) which has in itself an enormous substratum of past experience but no foundation of inescapable future. Such a proposition is not a certainty but rather a contingency embodying a query which, however incredible it may seem, may have a positive or negative answer; it is a fact, not a question of logic.

So long as a general proposition of the type given remains doubtful it is called a "hypothesis". A hypothesis, therefore, necessarily has an interrogative content, in so far as it awaits agreement or dissent. Consequently it differs radically from the "axiom" (discussed in the previous chapter), which is an assertive proposition to be accepted without requiring evidence or demonstration. If someone asks, "Is a straight line the shortest distance between two points?" he is formulating a hypothesis: but the mathematician, doing away with the question mark, gives this expression an axiomatic value with the seal of truth.

Other propositions which express a fact or condition are propounded as neither hypotheses nor axioms, but are asserted because demonstrable and demonstrated as true. In such a case the general assertion is called a "law"; and in certain other cases (notably those more complex and — according to some — not open to experiment) it is called a "theory".

For our immediate purpose these ideas can be presented only in a very general form. They will be taken up again later (section 542 *et seq.*). Some very common instances will suffice as examples.

The rhythmical rising of the sun is held as law by all men; the expansion of the universe is a theory; the carcinogenic action of cigarettes is a hypothesis; and the constant velocity of light is neither hypothesis nor law, but simply an axiom of relativity physics. Despite the apparently clear distinction, these propositions do not concern single events but express in general form something about the external world. The first two state a certainty or at least an opinion, the third expresses a doubt, and the fourth announces a condition to be admitted without onus of proof or evidence.

The accepted opinion is that law and hypothesis are statements about natural facts which will here, in general, be called "events" (section 354).

3131 Interpretative precautions

However, it can be seen from previous considerations (sections 172, 311) that things are not quite so simple. The only facts of nature are the "elements" and "sensations" of Mach, or in more modern terms, the "states of affairs" of Wittgenstein, or better still his "elementary propositions" (section 141). If someone says, "I see the colour red" or "I hear a noise," he is describing the above sensations; but if someone says "the toreador's cloth is red" and "the miaowing of a cat is a noise", he is formulating a "model of reality" (Wittgenstein's axiom 2.2). These are intuitive ideas, and yet one recognizes in them thought (axiom 3 of Wittgenstein), that is, the representation of states of affairs. The passage from sensation to thought is by an almost imperceptible shift.

A simple sensory fact sometimes enters into consciousness (in Locke's sense) already conceptualized. It can be inadvertently exchanged for a natural fact, and one should be on one's guard against this misunderstanding. The elastic which customarily held together the folder containing the pages of this manuscript was thrown on to the table and "spontaneously" formed itself into the graceful shape drawn below. First thoughts suggest that the event was determined by the will

Fig. 1

of the observer, or more correctly (though it amounts to the same thing) by the will of the manufacturer of the elastic, whose technical methods allowed the band to form these closed spirals.

When the same piece of elastic was thrown on to the table a number of times, it formed other figures, all graceful and all of the same type, which in a certain sense could be called equal because they are composed solely of interlacings of closed curves. Clearly, to say that the elastic behaves "naturally" is to leave out the manufacturer, to exchange for *res extensa* a "fact" composed of objectivity and subjectivity.

314   Heisenberg and the unification of objective and subjective

The definite success of the principles laid down by Mach and the denial of what Heisenberg has called "dogmatic realism" may be due to the formulation of quantum physics by the Copenhagen school, following Bohr[1]. From quantum physics it is clear that all the data we

---

[1] W. Heisenberg, *Physique et philosophie,* French edition, Albin Michel, Paris, 1961, pp. 81 *et seq.*

call "facts" are the result of an interaction between object and subject; that is, they are a description of Nature not as she is — *die Dinge an sich*, the metaphysical concept — but as revealed by the sense organs and by research methods for observing and describing her in clear language.

Einstein himself, who (as pointed out in section 261) laid great stress on the effectiveness of imagination in scientific research[1], always remained wedded to the illusion that what the mind foresees and experience confirms is an objective reality, is *res extensa*. But the electron, as previously said, is not a fact but an idea; nor are many concepts of quantum physics facts. Einstein did not adapt himself to this truth, and so never overcame his suspicions about the principles of the Copenhagen school. He said that the latter, while mathematically unexceptionable, offer no starting-point for further development. Here are his words: "It is my opinion that quantum theory of today represents, by means of certain well-established basic concepts largely derived from classical mechanics, an optimal formulation of interdependence. I think, however, that this theory offers no useful point of departure for future developments. On this point my expectation differs from that of the greater part of my physicist colleagues." And in another context, explaining his dissent, he added: "The justification of such constructs which for me represent reality rests only on their ability to make intelligible what the senses reveal."[2] That is, they are not valid as thoughts, but for the facts they contain.

3141   Dingle's comment

Assent and dissent over the theses of Mach and Heisenberg have today, however, become a question of proportion. This is because of the balance between the two components, natural and subjective. These vary with the temperaments of the scientists, with their branch of specialization, and with the phase of historical development.

---

[1] A. Einstein, *Conceptions scientifiques, morales et sociales*, French edition, Flammarion, Paris, 1952, p. 106. On science and imagination see K. Pearson's *The Grammar of Science*, pp. 31 *et seq.*: "The laws of science are products of the human mind rather than factors of the external world" (p. 36).

[2] These and other quotations are from the work *Albert Einstein, Philosopher Scientist*, The Library of Living Philosophers, Evanston, Ill., 1949. They have also been given in German and the original English and commentated by P. Caldirola and A. Loinger, "L'interpretazione della teoria quantistica", in *La Scuola in Azione*, 1959–60, No. 19.

In a memoir written in 1938, several times quoted (sections 1421, 171, 252), Dingle, disagreeing with Milne and Eddington, lamented that physicists were not well able to judge how much there was of the conceptual in quantum physics, and how little certain definitions of it could be considered factual. In 1961, however, when bringing the memoir up to date for an Italian edition, Dingle altered his previous point of view, and recognized that his earlier polemic position was no longer tenable because physicists of today are engaged in a thorough probing of the relationship between fact and theory[1].

The same thing has happened with the older biological sciences. These had previously collected an immense amount of documentation, and had built upon it the evolutionary systems of the nineteenth century (Lamarck, Darwin, Wallace, Häckel, Weismann, Nägeli, Rosa, de Vries, Montandon and others). Today these subjects are concerned with inquiry into facts, and include biochemistry, morphology, ultramicroscopic physiology, genetics, and so on. Each of these branches of biology as, for example, genetics — suggests and works out theories along its own lines, renewing and augmenting factual evidence.

32 Species and genera

We read in the first chapter of the Book of Genesis that "God created great whales, and every living creature that moveth, which the waters brought forth abundantly, after their kind, and every winged fowl after his kind" (verse 21), and likewise that "God made the beast of the earth after his kind, and cattle after their kind, and every thing that creepeth upon the earth after his kind" (verse 25). In the second chapter we read that when Adam[2] had seen them "he gave names to all cattle, and to the fowl of the air, and to every beast of the field" (verse 20); and that "whatsover Adam called every living creature, that was the name thereof" (verse 19).

A divine work, then, was the creation of animals and of things, but a human work, subjective, the naming of their categories; a perfect distinction between what man can know through revelation and probe into by metaphysics — such as the origin of the world or an act of faith — and that which concerns science, that is, the physical world co-ordinated and expressed in language.

---

[1] H. Dingle, "The rational and empirical elements in physics", in *Philosophy*, London, vol. 8, April 1938.

[2] In Hebrew, Adam is a common noun and means "man". Various peoples such as the African Bantus and the Eskimos call themselves by this same name.

The common noun is, in short, the linguistic expression by which we recognize the conventional similarity of groups of events, and the dissimilarity of these with respect to other groups with different names. When we say "dog" we know we are affirming the similarity of various breeds, from the tiny chihuahua to the giant St Bernard. But the difference between these breeds is much greater than between varieties of oxen, butterflies, crystals or electric sparks. The formation of the common noun will be discussed later (cf. especially section 356); for the moment it is enough to state its importance as a means of recognition, communication and action.

But the biblical text allows men the choice of names for living beings and also for things, and likewise their grading into "species" and "genera". From this there springs a further achievement worthy of note. In the same way that common nouns summarize similar experiences as analogous (dog, falcon, . . .), other classificatory words link names among which objective analysis recognizes analogous relations (mammals, birds, . . .). It is a question, then, of unifying from the bottom towards the top, from more restricted and better specified classes to more comprehensive ones, and this process can be pushed as far as one likes. Its success is a question of the capacity to abstract, and of practical necessity.

Consider the difference between the rough conceptualization arrived at by certain human groups, and that of peoples of high culture. According to Professor H. O. A. Wold of Uppsala University, the Eskimos of Greenland have about forty words to indicate different features of snow, such as dry, damp, recent, dirty, . . ., but no generic word for snow. Vice versa, in the West, everything has a name and is classified at different levels.

321 Classificatory paradigms

The common noun, which is an essential tool of knowledge and communication, assumes a greater operational interest in the hierarchical classification of events in "genera" and "species". The following cases are pointers:

(a) In a collection of Attic vases of the fifth century B.C., brought to light during excavations, numerous varieties have been recognized, classified by size, shape, purpose, decoration, etc. By patient selection it has been possible to form groups alike in some characteristics though not in others, but converging to a fairly identifiable type, to which has been given the name of *lekythoi*.

The lekythoi were originally vases for unguents and later became funeral furnishings. They are distinguished by a slender shape — the neck long and black, ending with a handle, and a cylindrical body with a white priming coat on which are line drawings of scenes, sometimes with a restrained use of colour. The class of lekythoi shows great variety, and since only a few names of artists and some places of origin are known (such as Timocrates or Sirisco), the vases are mostly distinguished by their designs. One speaks, consequently, of works by the painter of Achilles, of Icarus, of the torch, of Thanatos, of Charon, of the square, the phial, bath, cane-brake, of the women, the triglyphs, etc.[1]

All types can be arranged and subdivided as under:

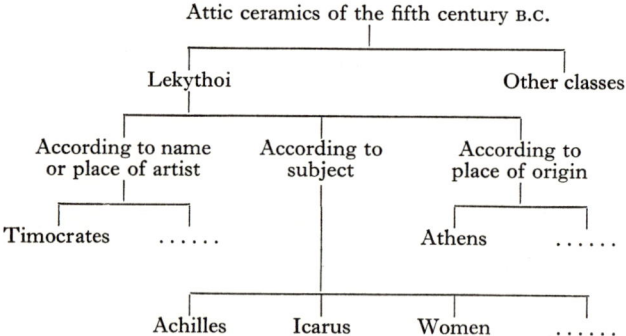

Notice that in the table all proper names are used as common nouns. Timocrates means "vases attributed to the artist of this name", and Achilles, Icarus, etc. mean "figures of these personages". Since the subject to be classified is Attic ceramics of the fifth century, even if subdivided into classes of small numbers, they would still be vases considered in the abstract, i.e. without individuating them.

(b) Among the tables used by naturalists the one opposite is useful as showing the distribution of the known types of sedimentary rock. It was drawn up by R. Fabiani (*Trattato di geologia*, p. 133).

---

[1] The vogue of lekythoi began about 480 B.C. and ended in 392, giving place to large vases of similar form in marble, with decoration in relief. The artistic quality of ceramic lekythoi is generally high and sometimes excellent — as in the case of the "painter of the women", to whom is attributed a very famous example. The prevalence of designs of women caused the malicious Aristophanes to say that the lekythoi are "collections of women". On lekythoi see the article by E. Paribeni in *Enciclopedia dell'arte antica*, vol. 4, Enciclopedia Italiana, Rome, 1961.

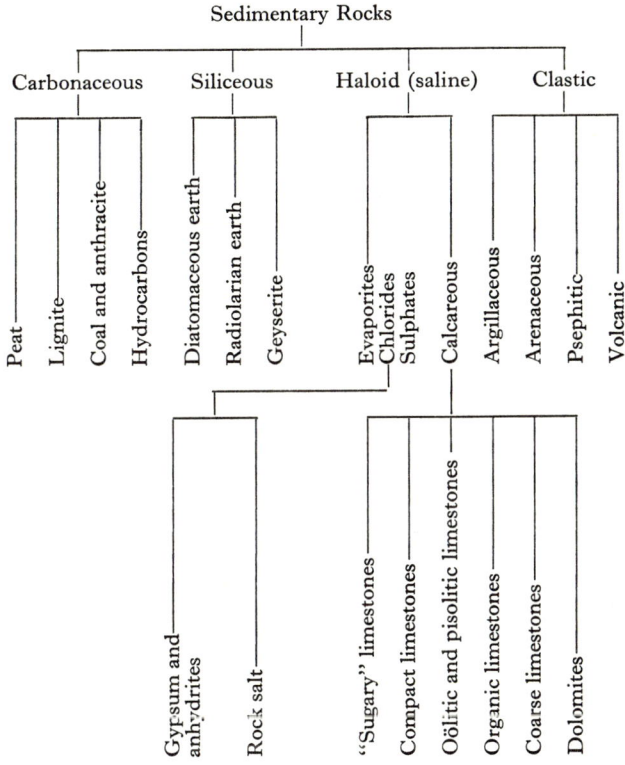

(c) The next table is no different in form from those preceding. On the basis of *tituli operum publicorum*, the names of those who commissioned public works in Ancient Rome can be classified as follows:[1]

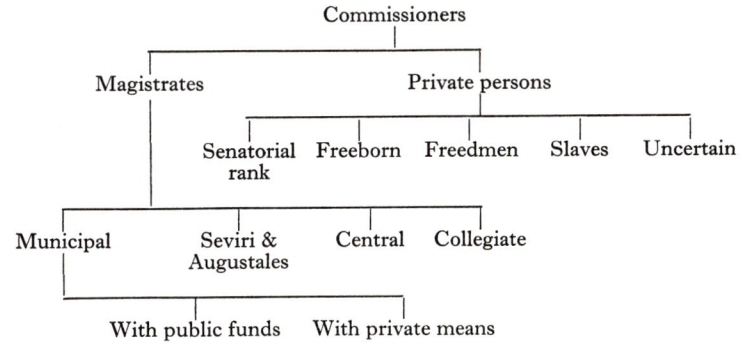

[1] See I. Calabi Limentani, *Studi sulla Società romana: il lavoro artistico*, Cisalpino, Milan–Varese, 1958.

It might be useful, but too lengthy, to pause and consider other classifications, contrasting one of the more detailed, such as the Linnaean classification of the organic world into genera and species with one of the many systematized ones proposed from the time of Comte onwards for the branches of scientific knowledge (section 58). The first classification (that of Linnaeus) starts with the dichotomy of animal and vegetable; the "animal" subdivides into vertebrates and invertebrates, i.e. from man to protozoa; the vegetable division includes the phanerogam and cryptogam plants. The first branch of these plants includes the more complex monocotyledons, such as orchids and palms, and the second the most elementary plants down to the blue algae and bacteria without chlorophyll.

Di Fenizio has differentiated the branches of knowledge into those which individuate (histories) and those which generalize (sciences), then further dividing this last group into even more specific and limited classes[1].

It is easy to see that all the above tables present similar characteristics (hence the possibility of a classification of classifications) and contain only common nouns. Each paradigm starts with one name at the top, increasing in number as it details towards the base.

Logicians call the most comprehensive class at the top of each table the *summum genus* and the numerous lowest classes the *infimae species*. It is not possible to go lower than these without ending up with single examples (e.g. the numbers of the catalogue of the lekythoi) and eventually with proper names which identify the examples (e.g. dedicators of public works). Each of the intermediate classes, between the top and the bottom, is called "genus" when considered as a grouping of the classes below, and "species" when considered as part of the class above which includes it.

322   Inductive and deductive reading of the paradigms

From the above classifications one can see, in the concrete sciences, the reciprocal relationship between analysis and synthesis, between deduction and induction. Interpreting any of the preceding tables, one can read them from below upward, going from species to genera, with increasing generalization, or from the top downward, from genera to species, with an increasing detailing of the ideas involved. The deductive character of this second procedure is evident; all the attributes of the

---

[1] F. di Fenizio, *Le leggi dell'economia, il metodo dell'economica politica e della politica economica*, L'Industria, Milan, 1958, p. 18.

species below are implicit in the *summum genus*, but they only emerge after analysis, that is, as one continues to read. This must be made clear, though it seems superfluous to emphasize the inductive or generalizing character of reading from below upwards, by which the common noun is formed.

These interpretations imply that in the inductive process the *infimae species* can be regarded as experimental data — and this seems incontestable, because there is nothing to prevent subdivision from reaching the elements of immediate observation. They also imply that in the deductive process the *summum genus* can be regarded either as an axiomatic proposition or as a hypothesis.

But what is true for each paradigm is also true for each section of it; each intermediate genus can be considered as the *summum genus* of the ones below, and each intermediate species above the lowest can, if necessary, be interpreted as an elementary fact.

Harrod and others before him (Braithwaite, Wisdom and myself) have recognized that in present-day studies "there is perceptible a tendency to overthrow the traditional dichotomy of deduction and induction in favour of a new canonical scheme constituted by the trinity of hypothesis, deduction, and verification. Within this totality deduction has an important role."[1]

In Chapter 1 we noted the existence of a concrete mathematics that develops deductively but has a factual content which could only be included by induction. Chapter 2 dealt with axioms of the concrete sciences, which, though not purely formal principles, could be developed like the axioms of a content-less logic. It has now been shown that every scientific paradigm can appear as an inductive or deductive table, according to the direction of reading it.

3221 The triad hypothesis–deduction–induction; the synthesis of water

So far, the interconnection of hypothesis, deduction and induction

---

[1] R. F. Harrod, *Foundations of Inductive Logic*, Macmillan, London, 1956, p. 5. The author, pointing out that we are dealing with a "modern theory" and a "new canonical scheme", adds that "deduction has been booming". Apropos of its supposed modernity, it should be pointed out that the use of hypothesis, deduction, and induction together in scientific processes was suggested and fully developed in the first printed edition (1942) of the work by M. Boldrini, *Statistica, teoria e metodi*, Giuffrè, Milan, 4th edition, reprinted 1962. In the present work I bring these studies up to date. Compare also R. B. Braithwaite, *Scientific Explanation*, a *Study of the Function of Theory, Probability and Law in Science*, Cambridge University Press, 1955, p. 12.

has been considered in the abstract. We now consider some practical examples.

Example 1   The experiment of producing an electric spark in a mixture of hydrogen and oxygen has been performed very many times; water is formed and at least one of the two gases disappears.

(a) There is the "highest-level" hypothesis — to use Braithwaite's term — where two atoms of H and one atom of O are necessary to form one molecule of water, and also — although it is not really the same thing — two molecules of H and one molecule of O are necessary to form two molecules of H_2O.

This hypothesis, it is true, has been suggested by experimental data (and implies an inductive process to be discussed later), but since no-one has ever seen three single atoms combine to form a molecule of water, one cannot speak of an actual fact, but only of a concept. If such a concept is expressed as a doubt, it remains a hypothesis; if it is held to be demonstrated, it is a law; if it is dogmatically asserted, it becomes an axiom, although by nature more specialized than the primary axioms of chemistry, such as the atomic theory, the molecular theory, and the law of definite proportions (stoichiometry, section 3553), from all of which it is derived.

Dwelling for a moment on the idea of hypothesis, we can deduce from (a) consequences which can be checked experimentally.

(b) From direct deduction arises a new hypothesis, on an "intermediate level" with respect to the previous one. It can be expressed as follows. According to the molecular theory (Avogadro's hypothesis) equal volumes of two different gases, at equal temperature and pressure, contain the same number of molecules. The highest-level hypothesis must be considered strengthened if, on passing an electric spark through a mixture of two parts by volume of H and one part by volume of O, both previously measured under identical conditions, water is formed without leaving a residue.

(c) But not even the intermediate hypothesis by itself can be verified experimentally without reference to Avogadro's theory. However, the experiment can be carried out, conforming to a series of "lowest-level" hypotheses of the following type:

an electric spark, passed through a mixture of 2 cc of H and 1 cc of O, forms water without a residue; or

an electric spark, passed through a mixture of 3 cc of H and 1·5 cc of O, forms water without a residue:

and so on.

It can be seen at once that the series of tests, if continued, would be infinite. And yet it could never prove the truth of the highest-level hypothesis, because this is stated in universal terms; because it is valid for atoms and not for molecules; because the introduction into the reasoning of Avogadro's hypothesis is arbitrary; because in its formulation the isotopes of hydrogen and oxygen have not been taken into account; and finally, because no logical argument guarantees its future validity.

(d) Nevertheless, the highest-level hypothesis is firmly accepted in science, that is, it counts as a law, and H_2O, the chemical formula of water, is its linguistic expression. Although not upheld by any logical principle, the conclusions reached have undoubtedly always produced results, and will presumably continue to do so; this is, no doubt, a comfort for workers in laboratories and in industry. It is particularly important to keep in mind that while deductive analysis is satisfactory, based on what, for simplicity, we will continue to call the highest-level hypothesis, the same cannot be said of the synthetic procedure by means of which that hypothesis was reached, starting from experience: hence the entire process is influenced by this pervading factor, induction — that is, by an element that cannot be eliminated and about whose interpretation there is no agreement.

3222 Interpretation through paradigms

The synthesis of water could easily have been expounded by analogy with the tables in section 321, putting the highest-level hypothesis as the *summum genus* and particularizing down to the *infimae species*, consisting of single experiments.

Paradigms of the above kind, and in particular the very complex Linnaean one (mostly due to Engler), are the result of experimentation, comparison, hypothetical generalization, acknowledgement of error, and retracing of steps. Today the definitions of living species are based on at least twenty characteristics, but in the *Systema Naturae* published by Linnaeus in 1735 the classification of plants was based essentially on their floral parts. Linnaeus, as a great naturalist, was perfectly well aware of this inadequacy. Only through the centuries-long labours of research workers has a system been reached, and this is still in process of being worked out (section 381).

3223 Econometrics

Example 2 The same logical position can be recognized in the most recent development of economics, which Ragnar Frisch has named "econometrics". Joseph Schumpeter, the noted economist, outlines the procedure as follows. Econometrics "involves the fundamental principle that construction of the theoretical set-up should *precede* the statistical work; the relations themselves are not suggested by statistical observations; they are postulates and not results".[1] In other words, experience leads to the writing of certain systems of equations; these are accepted as axioms and are tested by further experiment. Validation, however, rests on past experience and never reaches universality.

3224   A polemic of Galileo's

Example 3   A further example is provided by Galileo's famous *Saggiatore* on how to cook eggs by centrifugation. Opposing the scholastic argument of Father Orazio Grassi (who used the pseudonym of Lotario Sarsi), Galileo reasoned as follows: "If Sarsi wants me to believe that in Suida the Babylonians cooked eggs by swinging them rapidly round in a sling, I'll believe him. But I must say that the cause of this effect is very far from the one attributed to it, and to find the true cause I would argue thus: if we cannot get a result that at some time others have got, it must be that we are lacking something which was the cause of the other, successful, result and that, as we lack only one thing, this one thing must be the true cause. Now, we do not lack eggs, or slings, or strong men to swing them round, and yet they do not cook. On the contrary, if they were hot they would cool off quicker; and since the only thing we lack is being Babylonians, then being Babylonians is the cause of the eggs hardening, and not the friction of the air, which is what I wanted to prove. . . . What the reasons may be to which Seneca and others attributed this effect, which is false, I leave you to judge."

The line of reasoning here is easy to follow. It starts with a highest-level hypothesis, "eggs cook when they are rapidly swung round in a sling", and deduces the tests to be carried out and the reasoning to be followed in order to arrive at the conclusion, which is that the hypothesis is false.

But the example is instructive for two new points it brings out, namely, that a single negative experiment is sufficient to belie a hypothesis, and that millions of experiments are not sufficient to demonstrate its truth. It will be noted that the present and previous examples are concerned with a highest-level hypothesis which is "always" true or "always" false. We shall examine later the intermediate case which is

---

[1] J. A. Schumpeter, *History of Economic Analysis*, Allen and Unwin, London, 1955, p. 1163.

of the greatest importance in contemporary epistemology, i.e. where a foreseeable event may or may not happen.

The example also brings to notice that the highest-level hypothesis of Father Grassi remains false so long as the proofs are limited to the scale proposed by Galileo. It would be confirmed if it were replaced in modern terms as follows: "The temperature of a body (eggs or anything else) which moves in the air is the positive or negative algebraic result of the heat convected by the wind and the heat accumulated by friction". At high velocities, in fact, supposing the positive component to be greater than the negative, bodies get hot and then, as with meteorites and missiles, burn out as they fall to earth.

3225   Isochronism of the pendulum

Example 4   The isochronism of the pendulum is a knottier problem. All his life Galileo sought in vain for a mathematical basis for it, which he would have liked to adopt as a highest-level hypothesis.

It is now known that the movement of a pendulum is a simple harmonic motion. The periods are only approximately equal for small oscillations, but we can accept the approximation which says that the arc covered by the oscillating bob is of about the same length as the corresponding chord. Small swing in pendulums can be said to be constant, since the thousands of experiments done with innumerable clocks have never contradicted this. That conclusion, however, does not constitute a logical guarantee for the future (cf. section 3552).

323   Revaluing deduction in science

The various cases illustrated have been chosen so as to cover only one particular category of events. In fact, all these phenomena are of a type which can be called inexhaustible, in that observing millions of cases (the synthesis of water, the oscillations of a pendulum) would not ensure the absence of exceptions. We postpone to section 473 a discussion of other types of phenomena whose limited number allows, at least in theory, an inductive generalization to be checked against fact (e.g. the resistance to strain in a consignment of ropes, or to wear and tear in a stock of shoes, on the basis of examining samples) (sections 346 and 473).

Furthermore, the examples given are intended not so much to emphasize the universality of inductive generalization as to bring to the fore the basic function exercised in it by deduction. The highest- and intermediate-level hypotheses and the rules of transformation, which correspond, for the most part, to the tautologies of logic and mathematics, constitute the real starting-point for scientific conclusions. These are further tested and eventually accepted or rejected, according to whether or not they conform.

In research into data and natural events, the function of deduction is therefore fundamental; and in accepting it, some workers have gone so far as to regard deduction as the only suitable instrument to hand for investigation, and as being equally valid for concrete and abstract disciplines.

3231 Confirmation of the role of logic and mathematics in science

The essential role of deduction in scientific inquiry is further confirmed by the fact that the method now in increasing use in nearly all the natural sciences consists of processes built up step by step on logic and having a mathematical character (Chapter 5).

In this connection, let us recall what was said in sections 18 and 181: namely that mathematics and the numbers used in the natural sciences share the same rules and concepts, but that in the concrete sciences, formulae can be interpreted in terms of natural facts and can then be worked out and become capable of mathematical interpretation. And as, in mathematics, even when symbols and numbers correspond to concrete quantities, the only possible development is the deductive one, deduction is therefore confirmed as an essential procedure for both the tautological and the factual disciplines.

33 Induction

The general faith in highest-level hypotheses and in deductions as criteria of judgement about experimental results relies on a remote, if not always explicit, appeal to a certain regularity in Nature — a metaphysical element which we should try to remove if we wish to find a rational basis (not a dogmatic one) for induction. From the latter, as has been shown, classifications and common nouns are derived and hypotheses and scientific axioms are suggested.

331 Mathematical induction

To avoid misunderstanding, it is as well to be finished with "mathematical induction" once for all with the remark that by such name is intended the fifth proposition of Peano's logic (cf. section 15). This states that if a property is (a) possessed by zero, and (b) if possessed by any number is also possessed by the succeeding number, then the property is possessed by all the numbers.

Mathematical induction has nothing to do with logical induction in the natural sciences, and no further mention will be made of it in these pages.

332 Scientific induction, from Socrates to Galileo

Manuals of logic, going back to the origins, record that Aristotle in his *Metaphysics* ascribes to Socrates the merit of having systematically developed inductive reasoning. Without underestimating the priority of our first father Adam, we may perhaps concede this priority to Socrates, with whose teaching much of the development of Western philosophy is linked. But historians point out that the strict use of the principle of induction can be traced back to an oft-repeated syllogism of Aristotle: "Men, horses and mules have long lives; the animals without gall-bladder are men, horses and mules: therefore, all animals without gall-bladder have long lives". It is a syllogism of the second "quantity", which draws a general conclusion from two particular premises, but the generalization is only apparent because, having made a list of *all* the animals which live a long time and do not have a gall-bladder, one derives the statement that "to put it in a single phrase, all animals without gall-bladder have long lives". Judge it how you will; but what the logical basis of generalization was, Aristotle could not have said, and the question is still open today.

On this point, consider the position of Renaissance science. Leonardo da Vinci, recognized as the originator of many ideas, wrote that "where one of the mathematical sciences cannot be applied, or one that is not linked with the mathematical sciences, there is no certainty". Apart from the rather obscure style of a "man without letters", as he called himself, it would seem that here is a first appeal to that convenient *deus ex machina*, always denounced yet always smuggled in under the cloak of the achievements of the physical sciences. Leonardo apparently meant that our observations of the world become valid only when, in response to a mathematical scheme, they reveal a natural rationality, a universal idea underlying them.

But Galileo must have been much more conversant with the principles needed for generalizing his observations, and he begins by reformulating, almost in the same words as Bacon, the principle of cause: "cause is that which, when present, the effect follows, and when removed, the effect is removed". For Galileo, the concept of cause plainly includes the two-fold relation of the rational link and temporal succession. The rational link, which resisted the denials of the empiricists, is for Galileo a mathematical connection. It will be dealt with later.

It is fitting to leave the comment on this most famous utterance of Galileo's to a contemporary philosopher.

"The mechanical hypothesis is only the condition for an interpretation of natural reality according to the necessity of mathematical laws. These laws constitute the form of rational comprehensibility, and allow each of its elements to be expressed as a function of the others. Galileo expressed this point of view clearly by saying: 'Philosophy is written in this enormous book which stays open in front of our eyes (I mean the universe), but you cannot understand it if you do not first learn to know the characters in which it is written. It is written in mathematical language, and the characters are triangles, circles and other geometrical figures, without which it is humanly impossible to understand a word; without these it is like wandering hopelessly in a dark labyrinth.' The intuition which in Plato remained simple supposition, held in check either by the limited Greek knowledge of mathematics or by a philosophic outlook not yet completely rationalized, here flowed freely and acquired the value of a method of research."[1]

This extract is instructive because it shows how, after an interval of 300 years, Galilean thought, after going through the sieve of rationalism, illuminism, empiricism, criticism and positivism, can be presented today in an acceptable form. What, in fact, did Galileo say, and what have so many subsequent writers repeated? — simply that mathematics is inherent in Nature, and therefore the human mind has not to invent it, but only to discover it.

In Chapter 2 it was shown that in the course of the last few decades the viewpoint has greatly changed. In the past I would have endorsed the opinion of Banfi, but today we must recognize the arbitrariness of axioms and the tautological character of mathematics, and likewise admit that scientific uniformity cannot look for support in universal principles of causation and in the mathematical uniformity of Nature. It can no longer be discovered, but only formulated.

3321   Obstacles created by rationalism

The philosophy of Descartes and his followers contributed to the rationalization of Galileo's scientific position so as to reach, with Spinoza, the celebrated identification of *ordo rerum* and *ordo idearum*. Leibniz himself, who admitted a rational order in Nature, was opposed to this exaggeration, as appears in dissertation no. 33 of the *Monadologia*: "There are two kinds of truths: those of reason and those of fact. The

---

[1] A. Banfi, "Vita di Galileo Galilei", *La Cultura*, Milan–Rome, 1930, p. 161.

truths of reason are necessary and their opposite is impossible, those of fact are contingent and their opposite is possible. When a truth is necessary, its rationale can be found through analysis by resolving it into ideas and more simple truths, because one so arrives at first truths."[1]

3322 Hume's reaction

A vigorous reaction to the rationalistic conception of Nature was developed by Hume, with his radical criticism of the naturalistic concept of cause. Hume asserted that all objects of human research can be divided into the two classes of "relations of ideas and matters of fact", as opposed to Spinoza, who asserted their identity.

Concerning the relation of cause and effect, he adds: "I venture to assert as a general proposition, which admits of no exceptions, that the knowledge of this relation is not obtained in any case through *a priori* reasoning, but comes entirely from experience". And further on, concerning the conclusions derived from the concept of cause, he adds: "These two propositions are far from being identical: 'I have found that this definite object is always followed by that definite effect'; and 'I foresee that other objects which are similar in appearance will be followed by similar effects' ." After stating that no reasoning can permit an inference from one to the other of the two propositions, Hume declares that the onus of proof lies with whoever believes it. The subject is discussed at length in other parts of his famous *Enquiry*, and the conclusion is that the concept of cause (and hence the validity of predictions drawn from it) has a purely psychological basis. "When we say that one object is connected with another, we only mean that the two objects have acquired a connection in our thought.[2]"

Experience first, and persuasion after. This is illustrated by the well-known example of a billiard ball which hits another and sets it in motion. No mental impression exists here corresponding to the idea of "necessary connection" between before and after, between cause and effect, and therefore one must conclude that such a connection does not exist. Furthermore, according to Hume, no rational necessity inherent in Nature would justify the transition from such a connection, and from

[1] G. G. Leibniz, *Monadologia, vulgo: Principia Philosophiae seu Theses in gratiam Principis Eugenii conscriptae* (1714), edited by G. de Ruggiero, Laterza, Bari, 1957.

[2] D. Hume, *Two Enquiries: Concerning Human Understanding and concerning the Principles of Morals.* Many English editions.

succession in the past, to the notion of cause which would permit its extension into the future[1].

From the negation of the concept of cause, Hume derived a radical criticism of induction. According to him, it would be impossible to guarantee with certainty the truth of any synthetic statement whatever regarding things outside the present, or at least outside actual experience. Furthermore — to use modern terminology — to derive analytically the truth of the statement that A will be followed by B would be possible only if B were implied in A, that is to say, if A were a necessary indication of the presence of B, or again — more drastically — if A were B; and this contradicts the very idea of induction.

3323   Contribution of the Illuminati

The influence of the English empiricists and especially that of Hume on scientific epistemology in the eighteenth century was undoubtedly considerable. Geymonat mentions that Lavoisier, in formulating the principle of the conservation of mass (considered as weight), attributed to it a simple methodological value[2]; and only later was it built up into a metaphysical entity by the very people who denied metaphysics.

On the other hand, d'Alembert has reservations, and in an article called "Expérimental", in the *Grande Encyclopédie*[3], was forthright in his disapproval of "the abuse of calculus and of hypotheses in physics". By this he meant to pay homage "à ce qui doit être l'unique objet de la physique expérimentale, à ces phénomènes qui se multiplient à l'infini, sur la cause desquels le raisonnement ne peut pas nous aider". D'Alembert warns that the physicist "ne saurait trop les multiplier; plus il en aura recueilli, plus il sera près d'en voir l'union. Son objet doit être d'y mettre de l'ordre. . . . Qu'il se garde bien de vouloir rendre raison de ce qui lui échappe; qu'il se défie de cette fureur d'expliquer tout que Descartes a introduite dans la physique." With regard to physics, d'Alembert does not fail to declare himself "bien éloigné d'en proscrire

---

[1] A careful exposition and a modern reconstruction of Hume's thought, also in comparison with the subsequent thought of Kant, can be found in G. H. von Wright, *The Logical Problem of Induction*, Oxford, Blackwell, 1957.

[2] L. Geymonat, "Storia del pensiero scientifico" in *La Scuola in Azione*, 1960–61, No. 1.

[3] *The Encyclopédie of Diderot and d'Alembert, Selected Articles*, by John Lough, Cambridge University Press, 1954, p. 68. The article "Expérimental" carries the initials of d'Alembert.

cet esprit de conjecture qui . . . conduit quelque fois à des découvertes";
but only to conclude that "les explications dans un cours de physique
doivent être comme les réflections dans l'histoire, courtes, sages, fines,
amenées par les faits. . . ."

3324 *A priori* synthesis and Kant's law of Universal Causation

The Kantian critique cut short the conflict of ideas, but only for a
while. At the outset, Kant had agreed with Hume's thesis, but soon
dissociated himself from it. Hume, in denying any meaning to proof of
a causal link between successive events, had spoken of "before" and
"after" (section 3322)[1]. Kant pointed out that there are reversible
subjective befores and afters (someone who runs his eye over a printed
page from top to bottom and then from bottom to top inverts the order
of the sensations), but there are also irreversible befores and afters (as
is the case when someone follows the motion of a boat on a river). The
rules by which irreversible events are put into order would be the causal
laws. The invariables make up the real physical world of Nature: the
experience that one has of it is not formed by subjective sensations but
by a rational process, the "synthesis", which transforms the "material"
provided by those sensations into a valid "intersubjective" experience.
From this latter, Kant derived a series of rules to which the invariables
defining the real world have to conform. One of these rules is the "law
of Universal Causation", corresponding to the category of "causality".
"Everything that happens," wrote Kant, "presupposes something
which it follows according to a rule." And again: "All changes take
place according to the law of the connection between cause and effect".[2]

3325 Rejection of causation and return to Hume

However, against the view that categories drawn from synthetic
judgements are *a priori*, and hence that causation possesses permanent
truth, one can object that such a view is the result of past experience,
intercommunicable and irreversible in time, but whose validity elsewhere or in the future cannot be asserted. The transcendental deduction
of the law of Universal Causation is a synthetic, not an analytic proposition, and its truth is not demonstrable *a priori*. This law does not

---

[1] This account of the Kantian concept of cause is summarized from the book
of G. H. von Wright, quoted above.

[2] "Alles was geschieht . . . setzt etwas voraus, voraus es nach einer Regel
folgt"; "alle Veränderungen geschehen nach dem Gesetze der Verknüpfung
der Ursache und Wirkung".

hold good in time in general, but in time until today. Hence the unconditional validity of causality does not follow from this law nor,
consequently, does its continued existence in the future. The same can
be said of the other Kantian categories; and with regard to causation,
Kant does not invalidate Hume's position and does not justify induction. In the *Critique of Pure Reason*, Kant himself had upheld the
concepts referred to, but seems to have doubts about them in the
Critique of Judgement. He sees clearly the impossibility of deducing
(analytically) special laws of Nature from the law of Universal Causation, and realizes the necessity of other principles to establish true
induction. Thus he approaches principles such as the Uniformity of
Nature and Limited Independent Variations as starting-points for
establishing an inductive logic. He declares explicitly that these principles are not demonstrable *a priori* but are only subjective assumptions.
It is important to point out that Kant gathers them together under the
title of *lex parsimoniae*, of *lex continui in natura*, and other names which
would be specifications of a more general principle, "die Zweckmäs-
sigkeit der Natur" (section 336) — "the purposiveness of Nature".

3326   The positivist obsession of universal determinism

With the advent of positivism in the nineteenth century, Hume's
viewpoint was again challenged, and Comte considers the principle of
Lavoisier, already mentioned, as a law of Nature, and chemical affinity
as a property impossible to reduce to physical laws[1].

Dingle[2] has admirably set out what he calls the "obsession" of the
nineteenth century, that is, universal determinism.

"Professor Whitehead's condensed statement," he writes, "will
serve the purpose of setting it before us. Referring to Tennyson's *In
Memoriam*, he says: 'Tennyson goes to the heart of the difficulty. It is
the problem of mechanism which appals him.

"The stars," she whispers, "blindly run".

'This line expresses starkly the whole philosophic problem implicit in
the poem. Each molecule blindly runs. The human body is a collection
of molecules. Therefore, the human body blindly runs, and therefore
there can be no individual responsibility for the actions of the body.'

"I think," continues Dingle, "it is one of the most remarkable
examples of the human power of self-deception that this argument

---

[1] L. Geymonat, *op. cit.*
[2] H. Dingle, *op. cit.*, p. 160.

should ever appear other than absurd, for the whole practice of physicists, even in the nineteenth century, gave it the lie direct. They had found laws which correlated certain movements — mechanical laws. Other movements violated those laws, so they were released therefrom and different laws were formulated for them — magnetic laws. Still other movements obeyed neither mechanical nor magnetic laws; they were excused both and given a fresh set — electrical laws. A fourth set of movements — animal movements — obeyed none of these laws, and in this case no one succeeded in finding any laws which they did obey. And what happened? Instead of saying, 'There may be no such laws, and these movements may be free,' physicists and philosophers and everyone else refused to grant them even the freedom given to magnetic and electrical movements. Instead, they said, 'There is no escape from the conclusion that these movements, which obviously defy the laws of mechanics, are in complete subjection thereto.'

"The true answer to the determinist riddle is that there is no riddle to answer. The whole difference between mechanical movements and others is that the former obey certain laws and the latter do not; there is no other difference. If, then, mechanical movements are called blind because they can be predicted, movements which cannot be predicted must not be called blind. Molecules have nothing to do with the matter; they are conceptions introduced to correlate phenomena other than movements. What we are concerned with are the movements of pieces of matter like human bodies and stars, and we find that we can predict the movements of dead human bodies and stars, but not those of living human bodies. A child could have silenced Tennyson's Maid of Sorrow by reminding her that she was no star."

The quotation is somewhat lengthy but it clarifies the problem of the blindness or non-blindness of Nature, in contrast to the positivistic viewpoint of the nineteenth century.

3327 Evolutionism and materialism

Particular examples confirm Dingle's historical account and critical objections.

Darwin's doctrine, put forward as a hypothesis with his usual strict caution, was worked up into a metaphysical instrument and gradually distorted in numerous ways, used as a symbol of materialism, or adapted to serve preconceived ideas such as the "fundamental biogenetic" law. Haeckel announced this law after making some observations on the strength of which he maintained that ontogenesis, or in-

dividual evolution, recapitulated phylogenesis, almost as if Nature —
the mistress of mechanical disciplines — out of some mysterious
caprice wished to develop man, making him go through the successive
phases of protozoa, coelenterates, worms, cephalopods, chordates,
fish, and onward up to mammals and the primates. Sociologists, in their
turn, took over this doctrine to theorize on the linear and predetermined
development of human societies, thus giving rise to positivistic sociology
— that of Herbert Spencer being the best known; and even historians
of art, such as Reinach, pretended to see an almost linear descent to
Raphael and Rembrandt from the paintings and graffiti on the cave-
walls of Cantabria and the Dordogne.

Putting renewed emphasis on universal causation and the uniformity
of nature, says Harrod, writers of the nineteenth century returned to
Galileo, forgetting the teachings of Locke and Hume, according to
whom only experience has a positive value for induction. "If every
cause must have its unique effect and every event must have one of a
limited number of causes", writes Harrod, "it was held that by a
careful scrutiny of certain phenomena and by using methods of elimination, one could establish that an event B was certainly an effect of an
event A in a class of cases; then, by applying the principle of the
Uniformity of Nature, one could generalize and hold that an event
of character A would invariably be followed by an event of character
B."[1]

333   J. S. Mill on induction; Harrod's criticism

As we have seen, Hume had not taught this, and consequently, for
those who remembered it, the positivistic position was becoming difficult to justify, while remaining at least convenient and attractive:
hence the effort made by J. S. Mill, a man of his time, who only half-
tried to break loose from the difficulties which it involved. The solution proposed by him is still a matter of dispute among logicians.

Harrod is frankly critical on this point. He begins by affirming that
the postulates of uniformity and causation have never been defined in a
satisfactory way, and yet, if they were simply assumed, they would be
nothing more than "innate ideas" in the sense so deprecated by Locke.
Mill, adds Harrod, recognizes this, and tries to get out of the difficulty
by arguing that the Uniformity of Nature can itself be established by
inductive reasoning. "If," says Harrod, "we proceed with our observations and make generalizations by induction from them, we soon find

---

[1] R. F. Harrod, *op. cit.*, p. 49.

that Nature is uniform. Thus uniformity, he [Mill] argued, can be accepted as a basic axiom, since much experience, perhaps all experience, provides evidence for it. This type of argument is irretrievably circular. If no mode of induction can be found that is valid without the support of any uniformity axiom, then we have no grounds for those very generalizations which purport to provide the empirical evidence for a general uniformity of Nature. Therefore, if we are to make some progress out of our original state of nescience, which is due to our birth into the world without innate ideas, there must be some mode of induction that is independent of any uniformity axiom."[1]

However, Harrod concedes that data collected up to the present time make the following two opinions sound very plausible: experience has taught that uniformity is inherent in Nature; and in many cases where one seeks to learn from experience one assumes in advance the existence of uniformity. These two opinions, Harrod concludes, are compatible in a different way from that suggested by Mill — a way which leads to the justification of induction.

The problem will be discussed in later pages.

334 Braithwaite's justification of induction

Braithwaite's attitude is less rigid than Mill's[2]. We have seen that, given a highest-level hypothesis suggested by experience, it is possible to reduce any scientific reasoning to pure deduction (sections 3221 and 323). Now, Mill's need was to reach that highest-level hypothesis by induction. Braithwaite says that, in the same way as a deductive inference is justified by the intermediate steps which link the major premiss to the conclusion, the correct way to discover the validity of an inductive inference is to analyse it minutely, in the hope of finding a suitable "suppressed" major premiss and of identifying the intermediate stages. When these stages are clear, the question of validity can be sorted out. By this procedure induction is justified, because it becomes "quasi-deduction", starting from the suppressed premiss or premisses.

Everyone will see the analogy of Braithwaite's position to Mill's need to deduce from observation a general premiss as a suitable starting-point for deductive reasoning.

---

[1] R. F. Harrod, *op. cit.*, p. 11.
[2] R. B. Braithwaite, *op. cit.*, pp. 259 *et seq.*

335   Rejection of the principle of Uniformity of Nature

Harrod aims a sharper criticism against those philosophers who, after raising objections to the principle of Uniformity of Nature, re-introduce it implicitly under the changed form which Keynes called the "Principle of Indifference".[1] According to Harrod, Carnap made a very subtle use of it, accepting as equally plausible and with evident apriorism certain alternatives about which nothing is known. Carnap "does not seek a logical justification for the choice in the old-fashioned sense. . . . The justification, such as it is, consists partly of rather vague recommendations of seemliness and simplicity, and partly in that he hopes to make it yield a set of inductive principles in conformity with those in general use."[2] It is not possible to pursue this subject further without introducing concepts which are not discussed until the following chapter.

336   Rejection of the principle of simplicity in Nature

Modern writers, with good reason, put the principle of Uniformity of Nature near to that of its simplicity. The historical development of this principle is learnedly and ably set out and criticized by a modern epistemologist whom it is instructive to read at length, although he introduces concepts not yet clarified in these pages[3].

"It was once believed that Nature herself was simple. The discovery of laws in what had up till then appeared a chaos of phenomena, the Pythagorean tradition of reality as made up of numbers, the Demo-critean tradition of Nature as formed essentially of matter and primary qualities and of change as reduced to a simple redisposing of parts, the Aristotelian–Thomist tradition that 'God and Nature do nothing in vain', all lent weight to this belief.

"And we have Kepler maintaining that 'Nature loves simplicity and unity'; Galileo convinced that what is most simple is most plausibly true; Newton asserting in the first rule of his *Principia* that 'Nature takes pleasure in simplicity' and disdains 'the ostentation of super-fluous causes'; and Einstein declaring: 'Our experience up till now supports us in thinking that in Nature the ideal of mathematical simplicity is realized.'

---

[1] See J. M. Keynes, *A Treatise on Probability*, Macmillan, London, 1921 (reprinted 1952), pp. 81–2.

[2] R. F. Harrod, *op. cit.*, p. 21.

[3] A. Maros dell'Oro, *La teoria fisica*, Cedam, Padua, 1955, p. 77. Bibliographical references have been omitted from the above excerpt.

"As long as science held up a mirror to reality, or approximated to it in the chaos of sensations as the Democriteans believed, descriptions (laws) and explanations (theory) of phenomena appeared truer the simpler they were. Occam's razor was at work, Morgan's axiom being only a restatement of it. In observing Nature, the complicated was only the superficial, and in science, because of human shortcomings, it was only provisional. In this connection there is the famous case of Fresnel, who felt he ought to apologize if his theory of optics presented somewhat complicated formulae. Time would certainly have resolved these formulae into something simpler.

"This belief in the simplicity of Nature and of science, originating with the Greeks and re-emerging in the seventeenth century, continued to gain ground through the whole of the nineteenth century. With Kant, for example, simplicity is the fruit of the *a priori* elements by which we arrange sensations and phenomena; but for him also Nature, the *a priori* synthesis, is simple, and hence so is science which expresses it (section 3325). Doubts began with the notion of contingency and Bergsonism on the one hand, and with the crisis of the mechanistic view, which spoiled the Democritean tradition, on the other. But the breakdown came when Mach (section 3111), Poincaré and the neopositivists showed how intensely conventional science was. Simplicity belonged only to laws and theories, useful to our mental economy, but not to Nature."

This section, though anticipating later treatment (section 563), was introduced at this point to clarify the position, because many authors still accept, more or less consciously, the simplicity principle which is backed by a long and splendid tradition. Although persisting until recently and supported by Einstein, it can no longer be accepted when scientific dogmatism is rejected and science is being built on essentially conventional bases.

34 Jeffreys' theory of induction

We have just remarked that uniformity and simplicity in Nature are, at bottom, correlative ideas, because anyone who assumes that the notion of uniformity can be derived from examination of an external world must needs rely on the principle that a substantial unifiability underlies the varied and changing appearance of events, and that this unifiability corresponds to the transition from the complicated to the simple.

Restating Mill's position in different terms, Jeffreys has formulated a theory of induction on which we will now comment, although at the

D*

moment it cannot be expounded in a suitable manner since it uses concepts which have not yet been introduced into these pages[1].

Starting from convenient theorems, Jeffreys arrives at the deductive conclusion that, as the number of observations of an event increases, in certain hypothetical circumstances the probability of its being verified in a further trial tends to certainty. This concept was reached without reference to the concept of cause (and hence without postulating any necessity in Nature), but its scientific acceptability for the future is a question of fact, not of mathematics. Jeffreys says that the number of alternative hypotheses (called "laws" by him) which can make a forecast true is always finite (however large one imagines it to be); and that the unequal validity of the hypotheses can be expressed as so many proper fractions whose sum is necessarily equal to 1, that is to say, to a number expressing the certainty that one or other of the laws themselves is the correct one. Jeffreys calls this conclusion the "simplicity postulate", pointing out that the aforesaid fractions can be disposed in order of decreasing size, and that the transition from the greater to the less corresponds to increasingly complicated schemes. It follows logically that the validity of a forecast of a phenomenon would be greater, the simpler the principle on which it is founded.

It should be noted that this conclusion was already part of the epistemology of Galileo, but that he founded it, as was said earlier, on the postulate that Nature acts in simple ways. Given that this idea, at least in its simplest form, is now no longer acceptable, it is difficult to replace it with the too analogous assumption of Jeffreys. Every formulation of scientific regularity (such as the increasing certainty that an event will happen next time as forecast) is always made, says Jeffreys, by adapting a mathematical formula to the facts. For the most part, fresh observations which deviate from expectation require the formula to be re-adapted, which nearly always results in its further complication. But there are notable exceptions. Precisely because Einstein's theory of general relativity, far from creating complications, enabled the phenomena of light and gravity to be understood together in a much simpler form than previously, it aroused immediate interest and was accepted as true. The circumstances were similar for quantum theory, originated by Heisenberg, which was simpler than all others put forward to cover the same phenomena. If we wish to express numerically the greater credibility acquired by the aforesaid theories as compared with previous

---

[1] Sir Harold Jeffreys, *Scientific Inference*, Cambridge University Press, 2nd edn, 1957, pp. 34 *et seq.*

ones, we should conclude that the simpler explanations correspond to the greater fractions, as Jeffreys maintains. But this type of interpretation carries the problem out of the logical into the psychological field and increases the doubt that the criterion of truth of scientific forecasts can be based essentially on an abstract mathematical theory (section 552). Relativity and the quantum theory are not accepted as true by physicists because they are simple, but because they co-ordinate and help in understanding better than do other theories, the phenomena to which they apply. Jeffreys' postulate takes no account of this fact. Hence the perplexity aroused by a theory which, on the basis of abstract principles, attempts to give a logical foundation to induction, that is, to establish the rational foreseeableness of natural phenomena on the basis of deductive considerations extraneous to them.

Perhaps these same doubts led Harrod to reject Jeffreys' postulate of simplicity, reformulating it in different terms (which will be examined in another chapter — section 553), and so to develop an inductive theory of his own.

341 Harrod's theory of induction

In section 322 it was said that Harrod accepts the principle that science is built on a ternary schematic combination of hypothesis, deduction, and induction, and (as many others had done before him) he understands that hypothesis is often "the play of creative imagination; it is in this activity that scientific genius can show its cunning".[1] About such a striking and alogical perception there is clearly little to be said, while it remains for the ordinary researcher to analyse the remaining two phases, passing from the deductive to the inductive.

Harrod imagines a journey through a "continuity" by a nescient man; it may concern a strip of colour, an imaginary street, or something of that kind. The experimenter does not know the whole continuity, even less at what point of it he is, nor how much lies before him. The observer slowly gathers his impressions, supposedly uniform, and records them.

After the first impressions, he may take a chance and assume that, at least for a short spell, the continuity will persist. At length he will gain confidence, and while remaining cautious in forecasting only a small fraction of the continuity (expressed in terms of experience already gained), he will perceive more and more often that he has hit

---

[1] R. F. Harrod, *op. cit.*, p. 8; similarly M. Boldrini, *op. cit.*, p. 81 *et seq.*

the mark. He will, however, reach a point where the continuity ends and has to admit that his final forecast was wrong.

We are dealing here with a hypothetical case of progressive induction, based solely on the experimenter's experience when completely ignorant of the external world — the ignorance postulated by Locke.

342   Induction in continuous and discontinuous phenomena

The above example was intentionally given in vague terms, to cover the case where experimentation and extrapolation from it proceed in a continuous manner. But by introducing new models it is possible to transfer gradually from the continuous to the discontinuous, so as to bring the example nearer to a concrete formulation.

To begin with, consider an experiment with observations which, along a continuum, can be considered as discontinuous. A typist starts to use her machine; she does not know the length of the ribbon, nor what portion of it has previously been wound off from the right-hand bobbin on to the left[1]. From the moment t the operator counts the number x of her taps — taking them as expressions of the continuous length of the moving ribbon, and from time to time she forms the hypothesis that she will have enough ribbon at least for the next tap, that is, that there will not yet be need to press the little gadget which reverses its direction. Clearly, if the ribbon is very long, the fraction $x/(x+1)$ will become greater and she will express a growing hope that it will run for at least one more tap. Moreover, all the forecasts will be correct except the last one, which will be wrong for the end of the ribbon.

343   A scheme of measurable induction

With a different set-up one can pass from a finite continuity, such as the typewriter ribbon, to an infinite one.

A circular opaque tube is transparent in the section AB (Fig. 2). By means of a valve at B, which opens in one direction only, the little balls which the tube contains can circulate towards the right, and their colour, black or white, can be seen as they pass, one by one, from A to B. The balls are very numerous (in the diagram there are only 27 black and 3 white). The observer has no idea what the tube contains

---

[1] This example is mentioned in M. Boldrini, The Presidential Address presented to the 32nd Session of the International Statistical Institute, Tokyo, 1960, and published in *Revue de l'Institut International de Statistique*, vol. 28, No. 1 and 2.

nor of the colour of the balls, but notes them passing successively from *A* to *B*, and after seeing a black ball he forecasts another of the same colour, and so on indefinitely. If *x* black balls have already emerged and a further one is forecast, there are two possibilities, i.e. that the experimenter guesses correctly or not. The ratio $x/(x+1)$ expresses the proportion of the known past results to the total number of experiences, including the forecast one, and can be taken as a measure of the plausibility of the expectation. It approaches unity the longer the past experience and the more limited the extrapolation. In fact $px/(px+1) >$ $x/(x+1) > x(x+2)$, etc., when $p > 1$.

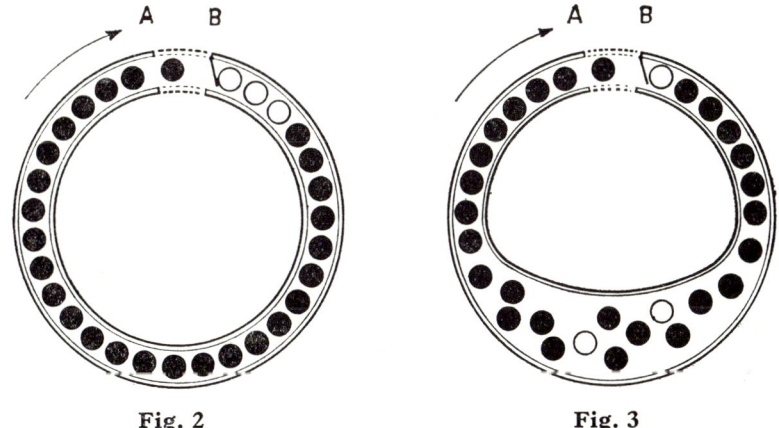

Fig. 2   Fig. 3

In Fig. 2 the last three balls are shown as white. Clearly, when they appear unexpectedly the forecast of another black ball will be wrong, and with the break in continuity the experiment could be considered as concluded.

However, this is not to say that real phenomena present themselves in this way. When, in midsummer, someone forecasts that the weather will continue fine on the morrow and is mistaken, he does not for this reason necessarily give up his favourable expectation, but for the day following the rain he again forecasts a fine day, with a reasonable prospect of success. This differing case is in line with the previous experiment, but generalizing the set-up as shown in Fig. 3. The black and the white balls roll from the point *A* to the point *B* inside the ring, but in doing so they get mixed up in the wider middle section. In this way their order ceases to be pre-established as it was in the preceding case, and the experiment can be continued indefinitely without the observer detecting any rhythm whatsoever, since new dispositions always occur. Further variations are possible by introducing balls of other colours.

344   Progressive confirmation by induction

These various experiments, when generalized so as to eliminate
discontinuity — that is, considering the taps on the typewriter or the
coloured balls as representing units of length along the ribbon or the
tube — would outline accurately what happens in Nature. Many
phenomena in the world flow without a break, and the observer distinguishes thereby some special characteristics (the ribbon shifting as
the letters are typed, the movement of the coloured balls through the
tube). At the start, the observer is in the state of complete nescience
postulated by Locke and, according to Hume, nothing can give him any
indication of the future development of the phenomena under observation. But gradually a conclusion is reached (for example, that everything will continue as at first for a certain space, or length of time, or
number corresponding to $1/x$ of the total already elapsed) and is expressed by a numerical quantity which measures the degree of conviction reached.

345   Harrod's principle of unconditional simple induction

Since the ratio $x/(x+1)$ increases as x increases in an objective way
(that is, without appealing to the principles of Uniformity of Nature or
of cause), wider generalizations can be inferred directly by simple
induction from the regularity observed in experiment, implying a
notable regularity in certain types of phenomena. This is Harrod's
conclusion, and he dwells lengthily on the subject, illustrating it by
comparisons.

A blindfold man playing blind man's buff on the carpet in a room
asks himself whether he will reach the edge of the carpet with the
next step. The event is not improbable, though if the forecast be limited
to one step, the edge may not be reached. If, however, instead of being
in the room, the man was set down blindfold at an unknown point in
the Sahara desert, his forecast of not reaching the edge in one or a few
steps — always bearing in mind the limitations of extrapolation —
would acquire a high degree of probability.

The presumed continuance of the desert and of the typewriter
ribbon are analogous to the foreseeable continuance of natural events —
that is, they are based on the objective measure of the expectation,
expressed, once more, by the ratio (usually increasing) $x/(x+\Delta x)$.

Harrod adds that, instead of a single continuity, one can imagine
having many and experimenting with all together. Then one must suppose that each forecast can fall from time to time on one of the continuities in any sector, taken to be as small as one likes. However it

may be, the expected value approaches nearer and nearer to unity without ever reaching it. As in Jeffreys' theory, unity is once again taken as the expression of certainty. Clearly, the case of many continuities is quite well exemplified in Fig. 3, where from time to time the balls are mixed up in an arbitrary way and hence determine different continuities following one another; but no rhythm can be recognized, and the experiment can be continued indefinitely as though there were a large number of balls.

Harrod calls the inference deduced from an experiment in a continuity and expressed by the relation $x/(x+\Delta x)$ the "Principle of Experience", or the "principle of unconditional simple induction". The principle becomes "conditional" when, dealing with many continuities, one must suppose, if only as a first approximation, that the experiments are equally distributed among the continuities (or over the area which represents them)[1]. According to this author, the "Principle of Experience", whether unconditional or conditional, would be the "objective" basis of all forecasts concerning natural events. To have observed that certain facts have up till now behaved in a certain way does not justify — as Hume had warned and contrary to what Jeffreys seems to believe — the logical deduction that they will continue to behave in the same way in future; but "nothing forbids one from thinking", adds Harrod, "that among all the possible manifestations the most plausible is a repetition. Discouraged by Hume", he declares, "logicians have tended not to face this issue fairly and squarely, but to run off in other directions", getting lost in vagueness or embroiled in contradictions and in begging the question, all of which things, in his opinion, could have been avoided.

346 Some comments

The link existing between the point of departure of Mill and that of arrival of Harrod is evident and admitted (cf. section 33). On the basis of experience the former derives a general affirmation of regularity, while the latter formulates from time to time an expectation which, on being verified, adds weight to the validity of the observations made. Mill's forecasts tended to become universal and hence independent of time: Harrod's are continuous and instantaneous and take place in time. It is here that their strength lies, but also their weakness. Whoever forecasts from past observation that the next oscillation of a pendulum will be isochronous with the previous one, and by oft-

---

[1] R. F. Harrod, *op. cit.*, Chapter 3.

repeated testing finds himself with a number of correct forecasts, is not doing anything logically different from someone who tries to forecast the more distant future relying on the regularity of Nature, because it is a question of guesswork with no logical justification. In a lottery, whoever backs the so-called losing numbers is no less deluded than the man who trusts in numbers dreamed of. Nor does a man behave rationally when he backs the ablest team in a football pool, even if his guess is correct. Concerning Nature, we only know that no one possible alternative is ever more likely than another, even if this conclusion — when applied, for example, to the rising of the sun — may seem contrary to commonsense. Harrod's faith, based not on long-term confirmation but on the Principle of Experience, could then engender psychological support, but his mathematical formulation certainly does not sustain it logically. This is the same sceptical conclusion which in section 34 had to be drawn from Jeffreys' doctrine.

Some writers have maintained that there should be a distinction between cases in which inference implies generalization to the infinite — as happens with isochronism — and cases in which inference is extended to a limited and, in the long run, verifiable number of instances, such as the life of a stock of electric-light bulbs determined by sampling (section 323). This is a partial concession to Harrod's position. The answer would seem to be that it is always a case of justifying experience in advance, and hence the problem remains unchanged, whether induction is applied to the first member of a limited class or to an infinity of possible future members[1]. But the question contains other implications which will be examined later (section 473 *et seq.*).

Consequently, the conceptual distance between Harrod and Mill lessens notably, and the progress realized by Harrod becomes an interesting sideline. The essential problem is not, perhaps, soluble in this way. Some of Harrod's remarks have a psychological appeal. One is as follows: "All that Nature contributes to the argument is a continuity. That is innocent enough in all conscience! It cannot be said that when we postulate that there is a continuity in Nature — from the dawn of experience we have myriads of examples of continuities — we are thereby postulating some inner necessity in Nature or making a prior assumption about Nature."[2] And again: "If Newton's Laws of Motion were *grossly* wrong . . . it is singularly curious that we have found no motions inconsistent with them through all this time and over so wide

---

[1] G. H. von Wright, *The Logical Problem of Induction* (quoted above), p. 3.
[2] R. F. Harrod, *op. cit.*, p. 53.

a range of observation."[1] One will agree with Harrod that "nothing prevents one from thinking" that these opinions are soundly based, but a man who barricaded himself behind a radical scepticism would require a convincing set of reasons to demonstrate them. Such conviction could neither be supplied by pure assertion nor by millions of past experiences. Forecasting from time to time, one can say with near-certainty that a pendulum will continue to oscillate isochronously; but Galileo's law still remains just as hypothetical and incapable of extrapolation, even as it does for Don Marzio in Goldoni's play, who had unlimited faith in his own watch.

347 Harrod's scheme of sampling induction

Simple induction — Harrod insists — is the only valid kind for forecasting future events. However, he formulates and discusses numerous other schemes, to which he attributes particular functions and tasks, which it is not possible adequately to summarize here.

It is useful to refer to what Harrod calls "induction by sample", without attributing to the word a strictly technical meaning. This type of induction is connected with the simplest and most elementary generalizations which men must have made, such as the opinion that the world around them consists of durable objects. This familiar conviction corresponds to statements of the following type. "If we shut our eyes and open them again, we shall find all as before; if we make bodily movements, the phenomena presented to us will undergo certain perspectival changes in accordance with a regular set of principles; if we walk forward in one direction, then turn through two right angles and walk forward again, we shall find the phenomena similar to those we have just experienced appearing in the reverse order."[2] Hume, Mach and Eddington had already described the familiar memory of their own writing-desks (cf. section 11). Harrod starts with his own study[3]: will it still be there (he asks himself) when I come back? And during my absences, by night or day, what happens in it? This is the ancient query of Protagoras — whether things vanish when they are not being observed. Harrod considers it too implausible that they should disappear (yet the *Mirabilis* opens its flowers when the sun is

---

[1] *Op. cit.*, p. 73.
[2] *Op. cit.*, p. 81.
[3] *Op. cit.*, pp. 82 *et seq.*; see also pp. 203 and 253. The same example for the same problem is given in H. Reichenbach, *op. cit.*, p. 177. The persistence of the table is also part of Harrod's theme, p. 203.

not shining on it!). A world of this kind, thinks Harrod, would be too complicated and unacceptable, without totally rejecting Occam's razor. On the other hand, it is only the absolute regularity of objects in reappearing when seen again, or alternatively their permanence, that allows us to reject the idea of a chaotic world like that of Heraclitus; which would be repugnant to our sense of order.

It is not the concern of science, but of the theory of knowledge and of metaphysics, to inquire what happens to reality when it is not observed. By the principle of indiscernibles, every real science remains indifferent as to whether experimental reality is permanent or renews itself identically each time it is perceived. But granted this, one may ask whether interrogatory propositions differ from questions about the future such as we discussed in the previous sections. It seems that they do not. There is in fact no substantial difference between someone who asks whether the next oscillation of a pendulum will be isochronous with those preceding — which he has, however, been observing at odd intervals, that is, by samples — and one who enters and re-enters his study (and we must assume that he does so at intervals) still expecting to find the table with his papers and the bookshelves with their contents. Now this type of induction has a distinctive characteristic — although differently stated — namely the time-variable which separates the certain experiences of the past from those expected in the future: and it is subject to the same doubts as simple induction.

35   Russell's theory of classes

There is a further inquiry of a more circumscribed kind which begins by defining a "class". The definition which Bertrand Russell has given is very general and serves well as an introduction to the subject, although Russell's classes are not the same thing as the generalizing classes which are of interest here and which, up till now, have been only intuitively anticipated (cf. section 32)[1].

According to Russell, a class can be expressed by enumeration of its members — or, accepting for the moment his nomenclature, "extensively"; alternatively by the synthetic formulation of some property of the class — or, as he says, "intensively". An example of the first category could be the subjects of the following enunciation: "Tiberius, Caligula, Claudius and Nero were the emperors of the Claudian *gens*", and of the second the enunciation "the Roman people".

---

[1] Bertrand Russell, *Introduction to Mathematical Philosophy*, 3rd edn, Allen and Unwin, London, 1924.

Enunciations of classes are reduced to "propositional functions" when they become the predicate of a variable. Thus, the above extensive enunciation would take the following functional form: "x is an Emperor of the Claudian *gens*". The function becomes true with the substitution for x of each one of the four above-mentioned names and becomes false with any other substitution, such as "Alaric" or "Betelgeuse" or "an electric spark". Conversely, the propositional function "x is a Visigoth king" becomes true not only with the name of Alaric, but with all the royal names of his race, Athaulf, Wallia, Theodoric, Torrismund, etc.; and the function "x is a star of the first magnitude" becomes true not only with the name Betelgeuse, but also with all the others which identify the most brilliant stars of the firmament, Aldebaran, Altair, Castor, Pollux, Sirius, etc.

From the intensive enunciation about the Roman people we have the propositional function "x is (or are) the Roman people", which becomes true for substitutions such as "Quirites" or "Romulidae" or others of that kind, and becomes false by putting "the Huns" or "the marsupials" in place of x.

All mathematical equations are propositional functions, because they become true for determined values of the variable.

Russell affirms that one approximates to a satisfactory (not a rigorous) theory of classes when one relates them to propositional functions, and he calls "classes" all the substitutions which make a propositional function true. Thus, in the propositional function "x is a Claudian emperor" the four names which make it true form a class, just as another class is composed of all the Visigoth kings, and yet another is composed of stars of the first magnitude with their twenty names.

351 Propositional functions

By the use of symbolic language, propositional functions can be expressed as ϕx and ψx.

If ϕx (or ψx) is sometimes true, this means that there exist arguments which satisfy it. Propositional functions constitute the only unexceptionable form of the predicate of existence; Russell says that all the other meanings are either derived from this or are a mental confusion. "We can say correctly 'men exist' only in the sense that 'x is a man' is sometimes true. But if we make a pseudo-syllogism: 'men exist, Socrates is a man, therefore Socrates exists', we are talking nonsense, given that 'Socrates' is not, as 'men' are, a mere indeterminate subject for a propositional function. The error is strictly analogous to that inherent in the following reasoning: 'men are numerous, Socrates is a man, therefore Socrates is numerous'."[1]

This argument elucidates the conclusion of section 14 concerning the predicate of existence, which (it was said) is not concerned with a definite subject such as "Joan of Arc", but is derived by the determination of the variable in the propositional function "x died on the funeral pyre".

It has been pointed out that classes are closely related to propositional functions but are not identical with them, because their existence is a "symbolic fiction". In other words one would say they are not facts but concepts — made not by Nature but by man.

352 Russell's classes and empirical classes

Among the conditions under which a symbol can serve as a class is that it should include the only arguments which make the corresponding propositional function true. Among other conditions, Russell proposes a very general and abstract one which allows the theory of classes to be used for that mathematical induction which was previously excluded from the scope of this study (section 331)[2]. It is important to point this out, because it justifies the reservation previously made that Russell's "classes" and "classes expressed by common nouns" are not identical concepts, but only nearly so.

Russell goes on to mention some propositional functions which imply the universal presence of a given subject, distinguishing these functions from those not implying it. He explains it thus: "Take an everyday remark such as 'A is a typical Frenchman'. How shall we define a 'typical Frenchman'? We can define him as 'someone who possesses all the qualities proper to the majority of Frenchmen'. But, short of paring down the meaning of 'all the qualities' to the point of excluding in practice *all* the qualities, we must observe that the majority of Frenchmen are not typical in this sense, and hence the definition shows that not to be typical is essential for a Frenchman."[3]

Here the difference between empirical classes and logical classes is accentuated: only the latter can imply universal subjects, while the former imply general subjects. Not only "a Frenchman", but also "an oak tree", "a rose", "a loved woman" can be typical entities, in the sense that they possess certain predetermined abstract characteristics (but not all) which justify the respective common nouns.

---

[1] Russell, *op. cit.*, p. 164.
[2] Russell, *ibid.*, p. 181.
[3] *Ibid.*, p. 189.

Russell's analysis is therefore important because with the concept of propositional function it makes clear the necessary summariness of the attributes which serve to define empirical classes. In saying that "x is a man-eater" or that "y is an oak tree", or that "z is a poem", one evidently accepts from the start that, to define classes, one never refers to a universality of attributes, because if this were so the common noun which serves as predicate would embrace a single subject, and the argument of the function would not be a variable but a constant. Furthermore, Russell's analysis distinguishes clearly between those values that make the function true and those that make it false, and hence makes comprehensible what are the common elements in the two sets of values.

3521 Propositional functions and the common noun

The position can be differently stated in the following terms. If all the determinants of the variable in a propositional function constituted a class, they could be assimilated to the species of which the predicate indicates the genus. It would seem, for example, that in saying "x is a man", one ought to mean that in the genus man are included all the meaningful substitutions of x. Otherwise, in the terms of logical analysis, the "meaningful substitutions" are proper nouns. "Socrates is a man", "Quirites are the Roman people" are explicit statements which ascribe a universal predicate to an individual, in contrast to the parallel, though different, branches of science which relate a class to a wider class, e.g. "a lekythos is a Greek vase", "the camel is a ruminant", "a planet is a celestial body" (sections 321, 4621).

The foregoing discussion is useful because with the propositional function it elucidates the step from the proper to the common noun, and with classification (section 322) the step from the common noun to the genus. In spite of these and other complications, many authors accept the common noun as the linguistic equivalent of a direct natural fact, and thus neglect to analyse the transition from the individual–universal, which is a matter of logic, to the class, which is the starting-point of every scientific process.

Carnap, however, did not overlook this, although he tackled the question from a decidedly abstract point of view. He[1] distinguishes "state-descriptions" from "structure-descriptions". The first are the combinations of predicates which qualify actual individuals. Individuals

---

[1] R. Carnap, *Logical Foundations of Probability*, University of Chicago Press, 1950, pp. 58 *et seq.*, 70, 108 *et seq.*

with a set of predicates in common (Carnap calls them "isomorphous" while in this book they are called "cases", section 46) can be grouped in "structural descriptions", which in these pages correspond to "events" designated by "common nouns".

If one identifies the "states of affairs" (*Sachverhalten*) dealt with by Wittgenstein in his axiom 2 with "events" designated by common nouns, and if one starts from the complex of events, one arrives at the idea of a world conceived as "all that which happens".[1]

353   Criticism of Hempel's observations

Naturally, the description of the world in Wittgenstein's logic has an analytic, not a synthetic character. It has been shown above that a classificatory paradigm can be read equally well from the bottom or the top (section 322), but that its genesis goes from the particular to the general and not vice versa. In a recent work of Hempel[2], it seems that insufficient attention is paid to the different ways in which the logical and factual sciences operate, or to the inductive nature of classification. Therein one reads that "the classification of the objects in a given domain D (e.g. numbers, plane figures, chemical compounds, galaxies, bacteria, human societies, etc.) is effected by setting up two or more criteria, so that every element d exactly satisfies one of them. Each criterion determines a class. . . ." The author exemplifies this concept through the well-known classification of the cephalic index according to five quantitative "criteria" as dolichocephalic, sub-dolichocephalic, mesocephalic, sub-brachycephalic and brachycephalic (section 382).

---

[1] The following axioms of the *Tractatus* should be reconsidered.

(1) Die Welt ist alles, was der Fall ist.

(1.2) Die Welt zerfällt in Tatsachen.

(2) Was der Fall ist, die Tatsache, ist das Bestehen von Sachverhalten.

(2.01) Der Sachverhalt ist die Verbindung von Gegenständen (Sachen, Dingen).

Using my own language, I interpret *Verbindung von Gegenständen* in the sense of "combination of events". Then the word *Sachverhalten* will mean "the classified cases which are indicated by common nouns", and the meaning of axioms 2 and 2.01 (and therefore of axiom 1) will become perfectly clear. "Every happening is the existence of events. Event is a combination of cases (entities, things)", and therefore "the world is all that happens, which is indeed formed by events" (1 and 1.2).

Translator's note. The first translation of the *Tractatus* contained some admitted errors. I have used a new translation by D. F. Pears and B. F. McGuiness, 1961, Routledge and Kegan Paul, London.

[2] C. G. Hempel, *Fundamentals of Concept Formation in Empirical Science*, University of Chicago Press, 1952, p. 51.

But this is no more than a confusion, whereby the whole problem of classification is put in a false perspective. Certain of the "objects" cited by Hempel can be subdivided by deduction only, albeit in mutually exclusive groups, while others cannot. Thus integers can be divided into even or odd, and the latter into primes and not-primes, etc. But such distinctions do not identify classes, although they are properties of the system of numbers, since one cannot define the idea of number starting from that of prime number, but only vice versa. To be convinced that these distinctions are not classes, that is, parts of a whole, it is enough to consider that the odd and the even numbers, the primes and the non-primes, are denumerably infinite — i.e., in Cantor's sense, are as numerous as the integers themselves.

But in moving on to natural events, one must distinguish. It is not that the "domains" named cephalic index, bacteria, etc. are divisible into mutually exclusive classes, but that they are themselves syntheses of classes with their own common nouns (e.g. dolichocephalic and brachycephalic men, aerobic and anaerobic bacteria). It is otherwise with certain other natural "events" instanced by Hempel, such as galaxies and chemical compounds — but only for contingent reasons. The Milky Way certainly appeared to men at first as a band of brightness, before they discovered a mass of stars in it, and doubtless chemical products were assessed by ancient philosophers on the basis of Aristotelian qualities and not by their actual appearance. Consequently, before the different natures of the celestial "objects" called the Milky Way, the nebula in Andromeda, the nebula in Canes Venatici, the nebula in Triangulum, etc., were identified, they must all have appeared quite similar and so were gathered into this great category of the galaxies. The present-day possibility of putting them in classes is due to newly-acquired knowledge (sections 321, 322) which integrates but does not change the inductive nature of the genus "galaxy". This digression should serve to emphasize that classification is concerned only with the facts of Nature, not with mathematical entities; and that given its generalizing character, it must be considered from the bottom to the top and not vice versa, that is to say, as going up from the *infimae species* to the *genera* and not descending from the latter to the former.

354 Generalization and the common noun

Locke said that our experience of the world is made up only of particulars; consequently, in communication between men, experience could not be expressed otherwise than by proper nouns. Such, for example, is the case with the entity known as the "Oak of Tasso" (the

oak in the shade of which the dying poet used to rest); or the famous
"Oak of Guernica"; or that particular oak which, for lack of a special
name, acquires one when designated by its spatial co-ordinates, or is
marked by the forester with a daub of paint for felling. But such a
picture, however exact, is clearly insufficient.

From now on, the substantive "entity" will be used in the technical
sense to refer to data and historical facts whose peculiarities are unique
and unrepeatable. Therefore the "Oak of Tasso" is an entity as are
also "Julius Caesar at the Rubicon", "the destruction of Hiroshima on
the 6th August 1945", "Tower Bridge", "Vostok III" and so on. It
is evident that a historical fact can consist of an elementary sensation
but also of a complex of sensations (which are entities) combined with
data and classificatory facts. As distinct from entities, these last are
always compound, although they can be formed by pluralities of
entities or classes of classes. From now on, the name of "events" will
be used interchangeably with that of "classes". History and the other
subjects dealing with individuals concern themselves with "entities,"
while science is occupied with "events". Entities and events, then, are
two linguistic expressions, each denoting the whole reality. And indeed,
the totality of historical facts on the one side and the totality of scientific
facts on the other, though each is composed of structurally different
elements, cover everything that happens — that is, the world, according to the first axiom of Wittgenstein (section 141: see also section
4347).

Commenting on a biblical text (section 32), we considered that in
gathering their first sensations, men must have passed from entities to
events, so that one cannot regard as exhaustive the idea that human
experience of the world is made up solely of particulars or, if you
prefer, of sensations.

Scientists have theorized over the transition from entities to events,
and a few examples may illustrate this. Consider oak trees again. Like
every other historical fact, they lose their individuality when defined
by certain attributes based on comparison and selection. And then, as
with Tasso's and the one at Guernica, so with others, one realizes that
the following "characteristics" converge: "It attains a height of 20–25
metres; the bark on the trunk is deeply grooved; the leaves, on short
stems, are sinuate-lobate and shiny on both surfaces. The flowers
develop at the same time as the leaves; the ones with stamens are long
and greenish-yellow; the one with pistils, 3 to 4 in number, are on a
stalk $2\frac{1}{2}$ cm long, which, however, lengthens considerably later, when
it bears the fruit. The latter, the acorns, are ovate achenes, at the base
of which is a cup-shaped cover." These are not all the attributes

which make the Oak of Tasso unique; but as soon as they are recognized as belonging to it, that particular object ceases ideally to be what it is, and becomes a "general fact", a "case" (section 3521), that is, an event characterized selectively, similar to all other individuals with the same attributes. In other words, it becomes an "oak tree", an example of *Quercus Robur*, an element of a "class" designated by this "common noun". We can then have propositional functions of the type: "The oak is a tree with a single perianth".

The same formative process holds good for many common nouns such as "choral symphonies", or "oscillations of the pendulum", or "H" and "O" or, finally, the "synthesis of water", and for all the others which are gradually introduced into common language and without which daily life and the very publishing of this book would become impossible[1].

---

[1] See the apologue "Un mondo impossibile" in M. Boldrini, *Zibaldone*, Cisalpino, Milan, 1947, p. 211, which emphasizes the impossibility of living felt by someone who finds he has lost the meaning of the classes of things and of sensations. Also, the short article "Il fatto unico" in the same volume, p. 173, brings to light the disorientation deriving from entities which do not repeat themselves and hence do not belong to general types.

The transition from historical data to classificatory fact could not be more impressively stated than in the little volume by F. Molina, *Le ricerche spaziali*, Editrice Studium, Rome, 1961, pp. 28–29. It quotes a passage from F. Nansen, *Farthest North*, Chapter VI, in which is expressed the wonder of a man at seeing for the first time the aurora borealis, one of the most impressive spectacles in Nature. "No-one can describe the glory which it presented to our eyes. The masses of fire were divided into brilliant ribbons of many colours, which intermingled and twisted across the sky to the north and south. The rays shone with the purest and most crystalline colours of the rainbow, especially violet-red, carmine and a very light green. ... It was an infinite phantasmagoria of brilliant colours, beyond anything imaginable."

There follows the cold reduction of this and of other amazing experiences into classes, designated by common nouns, which are here shown by inverted commas. "This description", writes Molina, "mentions some of the forms of an 'auroral phenomenon' and some of the shapes which an 'aurora' can take, such as 'ribbons', 'bands' and 'rays'. From these shapes several classifications have been made. Among the principal ones are 'bows' and 'bands'; very characteristic are the so-called 'flaming aurorae', etc."

Common nouns are used by Nansen to describe the uniqueness of "his" aurora borealis, an entity in the sense explained in the text; and Molina uses common nouns to qualify the characteristics of those "events" which science groups together as the "aurorae boreales".

355   Style and classification in art

What the critics call "style" is no different from classification, scientific or practical; and using signs and indications in attributing a work of art to a presumed author is none other than induction. Therefore, definition by style tends to lead back into the field of events, the peculiarities of those very complex entities, works of art. Try playing to a moderately musical audience one of Mendelssohn's *Songs without Words*, or get someone who likes poetry to listen to some of the *Petits poèmes en prose* by Baudelaire, or ask a virtuoso to look at one of the excellent still-life pictures by Chardin. Most will identify the author or painter, making some vague remark about his compositions. If you then ask how they decided, they will answer that "one feels . . ." or "one sees . . ." if they are average persons, or in almost incomprehensible jargon if they are professional critics. First Luigi Lanzi, who was originally a Jesuit art dilettante, and then Giovanni Morelli, who had been a doctor with a positivist education before becoming the leader of a generation of art critics (including Adolfo Venturi, Toesca, van Marle and lastly Berenson) and a most fortunate collector, understood very well that so-called "style", when closely analysed, consists in the almost automatic repetition by an artist of a number of minute details — the so-called "stylemes" — which are considerably more valid guides, because inimitable, than (for instance) the form itself, in deciding unknown or disputed origin. Morelli gave importance to the modelling of the lobe of the ear, the shape of the ball of the finger, of the nails and eyebrows, and to more complex points such as the movement of drapery, the selection of colours, and so on — a procedure and method almost infallible, since it opened the way to great discoveries by a perceptive mind like his. In the less rigorous climate of today, the method, although loaded with semi-technical jargon, still allows (for example) the distinction between the very similar statues by Giovanni Pisano and Tino di Camaino to be based more on the wide hair-parting and the gap between the big and first toes than on the general layout or on spoken or written tradition.

The method usually called "naturalistic", because stemming from remote tradition as found in Lanzi and Morelli and their near and distant followers and seen in nineteenth-century literary critics such as Campbell, Lutoslawski and others, and in the modern scientific experts who make use of X-rays and chemical analysis of colours used (knowing, for example, that the ancients lacked Prussian blue), or the history of words; the length of literary sentences; the choice and placing of adjectives; the frequency of vocal and consonantal phonemes (which is so closely connected with softness or harshness of expression), and so on — this method continually returns to favour, unchanged in essentials, and elucidates the concept of style which, to anyone who wishes to understand something of it and is not satisfied with airy descriptions and paraphrases, turns out to be simply an explanation of classificatory attributes and could be summarized by common nouns. For example, Beethoven is associated with so-called musical development, Wagner with modulations; and the classical painter of exaggerated human figures is unhesitatingly Domenico Theotocopulos, called "el Greco".[1] Thus, with respect to the development of musical phrases there is the class of Beethoven's sonatas, and with respect to the disproportionate length of the limbs the class of el Greco's paintings, just as there are oaks and isochronous pendulums.

3551 Leonardo's clouds and Rorschach's ink-blots

To see how the formation of common nouns is a particularly flexible process, one may instance the pictures which Leonardo da Vinci saw in the clouds, and the objects which psychiatric patients see in the ink-blots of Rorschach[2]. Both the clouds and the blots are purely random, hence unrepeated and unrepeatable, worthy of individual names. And yet human fantasy arranges them and makes the different seem alike, so as to create a class to which Leonardo gave the common attribute of "human figures" visible in the clouds, except that he individualized them afresh in his drawings.

Would one therefore suggest that the world itself is just as random? Certainly not, since we undoubtedly have experience of facts which are unique — that is, unclassifiable — and of very common facts whose repetition is expected. But neither are "unique" facts ever truly unique: a very rare illness may show symptoms such as fever, tachycardia, hypertension, haemorrhage, oedema, etc., that are common to many other illnesses; nor do those diseases that can attack a person more than once ever recur unchanged.

---

[1] I have frequently returned to this theme, feeling able to take a more radical view (risking professional disdain thereby) after gaining a clearer insight into the close relation between naturalistic analysis and the analysis of art, and into the unity of the epistemological evaluation of natural facts. See, for example, the following monographs: M. Boldrini, *Statistiche letterarie ed altri saggi*, Vita e Pensiero, Milan, 1948; *Zibaldone*, p. 165; *Alla ricerca del tempo nell'arte*, Mondadori, Milan, pp. 29–30.

[2] Cf. M. Boldrini, Presidential Address, *loc. cit.* For a deeper rationalization proposed by Wittgenstein, cf. section 5643.

3552 Theoretical and experimental pendulum

Another example. The common noun "isochronism of the pendulum" has been mentioned several times (section 3225). What corresponds to this linguistic fact? The "simple pendulum" as defined by physicists is an abstraction; they are thinking of a heavy particle made to move without friction in a smooth circular arc in a vertical plane. Clearly such a set-up does not exist. In the laboratory it is replaced by a little ball suspended by a thread (the weight of which is considered negligible compared to that of the ball), which is caused to oscillate. But in this experimental apparatus, the approximate accuracy of the instrument, and the time factor which makes successive oscillations heterogeneous; the limiting condition that these are small; the fact that a sinusoidal movement, subject to the law of acceleration, is treated as if it were uniform; the approximation that the length of the arc travelled by the moving object is identical with that of the chord of the arc itself; and finally, as though that were not enough, the internal motion of the electrons which (as with Eddington's work-table — cf. section 11) constitute the ball: all these factors are left out of consideration in defining the isochronism of the pendulum. Instead, an approximating property is attributed to it, always verified in the past and supposedly indefinitely repeatable in the future. The phenomenon called by the common noun "isochronism" is therefore an inductive abstraction, not an incontestable fact: a creation of the mind, not a natural datum[1].

3553 Theoretical and experimental ideas of water

But let us return to the very significant case of the synthesis of water, proposed as a highest-level hypothesis in section 3221. In Italian idiom it is said that two perfectly similar things are as alike as two drops of water. Up to a few decades ago this held good, but since the mass spectrograph has revealed the existence of isotopes, things have become vastly more complicated. The isotopes of hydrogen have increased in number since the first discoveries and, to ordinary hydrogen, with atomic number 1 and atomic weight 1, have been added successively deuterium with atomic weight 2, tritium with atomic weight 3, and finally, in 1962, hydrogen H_4 with atomic weight 4. There are also three isotopes of oxygen with the atomic number 16 and with respective atomic weights of 16, 17 and 18.

---

[1] On the individuality of the real oscillations of a pendulum, in contrast with the uniformity of those of an idealized pendulum, see E. Mach, *op. cit.*, p. 272.

Therefore a molecule of water, with the formula H_2O, which was of unique structure up to the discovery of isotopes, has come to acquire ever more numerous isomorphs up to the thirty possible today (corresponding to the 10 binary combinations of two atoms of the four isotopes of hydrogen with one of the three isotopes of oxygen)[1]. But the number of molecules in a drop of water is a little less than two followed by 21 noughts. It follows that, however rare are deuterium, tritium and H_4 in hydrogen and atoms with weight 17 and 18 in oxygen, there are almost certainly present all thirty molecular combinations in every drop of water. Until fairly recently it was held that the percentage composition of H and O, as far as concerns isotopes, was more or less constant, but in 1935 it was discovered that the isotopic composition of oxygen in the air and in the water of Lake Michigan was unequal. Hence a new factor useful for further differentiating drops of water. But there is still another, deriving from the recent discovery that the law of simple and constant proportions, accepted by science for more than a hundred years as the basic criterion of a chemical compound, is valid only for a molecule taken singly; while the distribution of the forces between adjacent molecules, at least in the solid state, does not allow (as had always been thought) of stoichiometric constancy, that is, the maintenance of absolute regularity of structure and composition.

Chemical water, then, remains an abstraction with respect to natural water with its free ions, its gases and dissolved salts, and the impurities it contains.

3554 Water and waters

It follows from all this that the atomic concept of one "water", and one only, was not an objective truth but, as shown by subsequent discoveries, a provisional abstraction. Thus the inductive concept "water" should not have been retained other than as an axiomatic hypothesis, suitable for deductive reasoning in practice and in science; as such, it is still quite satisfactory. In terms of average values — because the natural mixture of the isotopes of hydrogen has a mass slightly greater than 1, and that of the isotopes of oxygen a mass slightly greater than 16 — the synthesis of water can still be described according to the plan set out in section 3221.

But these observations show that actual reality is somewhat different. Water, as once thought of, no longer exists, but there are infinitely many waters, and the resemblance of two drops is much less tenable

---

[1] M. Boldrini, *Statistica, teoria e metodi*, Giuffrè, Milan, 1962, pp. 97 and 426.

today than in the past. Perhaps tomorrow, through the discovery of
new isotopes, for example, or of the possible importance of the dis-
position in the molecule of two different isotopes of hydrogen, the
situation could change again.

3555   Subjective notion of water

Once "water" is thus reduced to a changeable scientific notion, it
is no longer a "thing" but becomes an idea, an event, a system of tiles
in the complicated geometric mosaic with which man makes and un-
makes his picture of the world; a common noun such as "oak", which
includes many varieties, historically defined and classified by their
characteristics. Can this mean at least that with the successive notions
of water which have held sway following the discovery of isotopes and
of non-stoichiometry, the chemical idea of water has increasingly
approximated to water as it exists in Nature? One tends to think that
although the real formula may perhaps be still remote from the present-
day notion, it exists objectively, and the purpose of science is to determine it exactly. Now, this idea is simply a prejudice, because science
does not in fact, postulate an objective "water in itself", which would
be the final aim and the end of curiosity about it. For science, there
exists only water which the atomic theory at present accepts and which,
after many vicissitudes, has supplanted the picture of it as one of the
Aristotelian elements. It is an ingredient of the present-day picture
of Nature, and the progressive notions of it have aimed only at presenting that picture as consistent. Water, then, is and will always remain
a concept, and if language has not followed all its variations (remember
heavy water and oxygenated water), the common noun which desig-
nates it at present corresponds to a scientific reality very different from
that current from the time of Thales and of Aristotle onwards.

This theme will be considered further (section 5512).

356   Repeatability and foreseeableness of classified events

The formation of common nouns — the examples in section 32
elucidate the process — passes through characteristic phases, which
might be designated as follows:

(1) Evidence of an approximate similarity between certain natural
facts and of a difference between certain others; for example, not only
does one feel intuitively a more apparent similarity of form between a
young sapling and a majestic centenarian oak than between this latter
and, say, a zoological or mineral specimen, but also as compared with
other species of plants, including tall-growing ones.

(2) Considered analysis of the intuited elements of similarity; in the case of the oak tree those attributes were chosen which allowed the quick botanical description given above.

(3) Attribution of essentiality to the descriptive characteristics of point (2) and ignoring, for the purpose of classification, all other possible differences. In the case of the oak, the actual size, foliation, position, and all the other numerous peculiarities which would enable one exemplar of the species to be distinguished from any other were eliminated from the list of "similar" elements. Take a characteristic which helps to define the species, e.g. that oak leaves are "sinuate-lobate". As the lobes can be few or many, botanists have not included their number in the attributes that identify the species. Similar observations could be repeated for all the other characteristics, both specific and non-specific.

(4) From the above operations emerges the common noun, which is taken as the subject of a propositional function in the predicate of which are summarized the characteristics of the species. One can for example say: "the oak is a tree with a single perianth" (section 354).

All this goes to confirm that "oak" (like "water") is not a material fact but a concept; to return to an earlier picture, it is a mosaic of facts and ideas with which the ordinary man and the expert compose a picture of this specific manifestation of Nature.

And then, with this mental build-up, a further point takes on value.

(5) Once a class — that is, an event — has been defined, we can deduce the possibility that this will continue to manifest itself successively; e.g. by planting an acorn, tomorrow just as yesterday, there will grow from it a tree which will repeat the definitive characteristics of the parent plant.

This is the problem of futurity: but of a futurity prearranged by man, with his schematic classification.

3561 The future in induction

The logical nature of this subjective future is incontestable. The constitution of a class or, if you like, the definition of an event, on the basis of the process outlined above, is a simplified picture of historical beings, a kind of model by which it is possible to compare other experiences as they occur. On the basis of the reasoning in section 3221 *et seq.*, the class becomes a highest-level hypothesis A from which can be deduced all the implications contained in it. Among these is the low-level hypothesis that, in foreseen circumstances, an event B will appear. The analytical reasoning will sometimes be simple and sometimes complex; however, its tautological nature confers a rational justification on it. Naturally, for B to be implicit in A, it must be part of the definition of the latter, but it could be consequential, or a particular case, or some other derivative form. In mathematics there are analogies which suggest intuitively these eventualities. Thus the confidence that experience will confirm the deductions made is based on the same process of formation as that for classes. This is not an historical fact, an autonomous manifestation of Nature, as the old epistemologies asserted, but a creation of man, a linguistic expression, a common noun. Time and space, in which the class is immersed, are not the "then" and "elsewhere" of historical beings, but attributes which are inherent in the class and which participate in its artificial fixity. This is so in daily experience as in scientific inquiry, for which reason it was thought necessary, in the first chapter of this study, to adhere to an interpretation of the world and of science in which language is, in fact, the final manifestation by which the behaviour of such a world is perpetuated and made communicable and intelligible.

Future experience might falsify the induction contained in the highest-level hypothesis or the deductions tautologically drawn from it. In such a case the process would have to be repeated and the whole scheme reformulated. But let us be clear on this point. The highest-level hypothesis is a classificatory generalization of the past and, as such, is exact. Therefore a denial of it can only lead to the necessity for a different schematization, for another interpretation, for a different framing of the historical entities on which it is based, for another form to express it.

On the other hand, the idea of inductive forecasting, as generally formulated, tacitly presupposes the recognition of a natural system which could be either regular — and hence confirm predictions on its development — or variable, and hence discrepant with its past. But inductive prediction as considered in these pages is a very different thing, and consists of the expanding of a mosaic made with standard tesserae and expressed linguistically, whence it receives the stamp of truth not from the external world but from its supposed pattern, that is (metaphors apart) from the context of the ideas into which it is inserted.

This being so, the only element which can change is not the external world but this context, in relation to which the inductive schematization adopted as highest-level hypothesis and its implications would become inadequate. In itself, the classificatory event, expressed by the common noun, lies outside historical time and space and would remain fixed if it were not always dynamically connected with the intellectual system of which it is a part. The pure rationalists overlooked this (and that was their limitation) and did not concern themselves with the need for continual control, but were satisfied even in science with the postulated identity of *ordo idearum* and *ordo rerum*.

It is customary to say that science is the search for regularity, but it would be better to say that it is the search for consistency. Consequently, scientific effort is directed not so much to an even better definition of what persists, as to establishing what is new. The real scientific discoveries are the novelties of science, the things not confirmed by previous induction, the doubting of highest-level hypotheses, and not the satisfied contemplation of their validity (section 5412). When one saw — or rather, when one had to admit — that living species change, first the notion of stability, and then that of biological evolution, were profoundly upset; and when it became possible to define chemical elements and then their isotopes, the notion of water had to be modified several times (sections 3553, 3555).

One arrives at the conclusion that, resulting from classification and deductive elaboration, common nouns correspond to sensations and stable events, and hence that they are foreseeable in their own space and time as long as their definition remains valid. The schemes of section 341 *et seq.*, all based on Harrod's inductive model, are nevertheless, still to be accepted, but only in the sense that, considered not as entities but rather as events described by common nouns, they lose the uniqueness through which they would remain immersed in the flow of history, and become definite linguistic expressions whose future is engrafted on their unchangeable past, and hence has complete deductive validity. If Protagoras' problem (section 347) about the permanence of objects when they are not observed is transformed from a description of an entity into that of an event, the problem ceases to imply a doubt but points to a certainty consistent with all that one knows and believes.

3562 Tautological character of the inductive future

The force of the foregoing notions is shown when they are subjected to logical or factual checks. They even reveal that Hume's rhetorical question as to whether the sun will rise tomorrow can receive a rigorous answer. Not because Nature is regular, nor because lengthy past experience gives certainty, but it is certain — failing proof to the concontrary — that the sun will rise, because this conclusion derives from the consistence of a scientifically accepted cosmological system, which explains the phenomenon. The regular rising of the sun is a tauto-

E

logical conclusion implicit in that system, as it would be if one were dealing with a mathematical theorem. The only difference consists in this, that as the cosmology is not an abstract axiom, but a combination of inductive generalizations based on the past, it possesses a universal validity only until the contrary is proved. In fact, if tomorrow the sun really did not rise, we would have to replace the whole system of ideas which implies this phenomenon as a necessity by a new general picture, adapted to justify the fact that from time to time the sun may not rise.

Cases more in line with the history of science abound and demonstrate that the non-occurrence of deductions drawn from inductive generalities already accepted as highest-level hypotheses compels theories and systems to be modified. Biological variations suggest cases in point. In fact, the discovery of new genetic mutations has often determined radical changes in previous theories of heredity, which began with the recognition that heredity was linked with sexual chromosomes and influenced by "crossing over", and later arrived at the modern concepts of the interrelation of nucleus and cytoplasm. The changing character and increasing complexity of the clinical picture explaining the emergence of many pathological forms may be adduced in support.

36   Induction as repeatability

In conclusion, induction is concerned with the problems of the future. This is in a sense an artificial future because the common noun which affirms it relates only to those elements which are conceptually repeatable[1].

---

[1] Maintaining that art is never realistic, I have written elsewhere as follows: "A man is a close intermingling of relations with the cosmos in which he is immersed and of which he is part. The epistemologists and philosophers of science know this well. By the exchange of oxygen and carbon dioxide he is blended with the atmosphere and, more indirectly, with sunlight which activates the chlorophyll in plants. These, together with carbon, atmospheric oxygen and water, build up carbohydrates, while by using the nitrogen in the earth they produce proteins. In eating these, herbivorous animals transform them into higher-grade proteins, a valuable part of human nutriment. Thus, together with many other substances, the same atoms of carbon which man breathes out, return to him. But our individual existence has genetic links with other men, going back into the past through the immense line of forebears, and into the future by the union of male and female. And that is not all: beyond material relations spring up sympathies and aversions, thoughts and feelings, sordid interests and generous impulses. And higher still comes that striving towards the infinite, which is in each one of us, and which God blesses and rewards with his promises and gifts. This is the reality of man, which can scarcely be approached in words and which the most objective of the masters of colour have never tried to paint."

The position adopted by Harrod, of arriving at induction by small steps, and that reached above, of attaining it by way of classified entities, represent different approaches. The one relies on the conformity of the future with the past, the other recognizes in conformity a conceptual predetermination. Carefully collected, arranged and schematized (so long as experience does not belie them), the conformities will form the basis of practical conduct for the common man and the hypothetical starting-point for scientific deductions; not an hour longer, not a metre outside the field in which they have had or now have value.

The position thus reached satisfies Locke's requirement of not invoking any innate idea (sections 3327 and 333). Harrod's criticism of Mill's methods of elimination is that they are not acceptable as useful axioms for separating the constant from the variable in defining things (and hence also in the forming of common nouns); and that they assume an extraneous principle that some things in the world are uniform and some are not. The system of reasoning we have worked out here avoids the objection that the formation of the "class" is vitiated by a misleading use of the ideas of cause and of uniformity of Nature.

361 Schlick on induction

The better to appreciate the previous conclusions, let us consider induction according to Schlick, who said: "Induction is not a logical process and it is not possible to justify it logically." According to Barone's summary[1], Schlick proceeded to reject "Reichenbach's attempted explanation based on probabilistic logic and on the introduction of values intermediate between 'true' and 'false'." Barone goes on to point out that Schlick "accepts the compelling necessity for Reichenbach's attempt: he recognizes the inductive character of science, which is not a simple description of fact, but a comprehensive and far-seeing effort towards action. The problem of induction consists in justifying logically the general propositions on reality, which are always extrapolations from single observations.

"In company with Hume, we recognize that for these propositions there is no logical justification, and there can be none because they are not authentic propositions. Natural laws are not (in the language of the logician) *general implications*, because they cannot be verified in every

---

(M. Boldrini, *Alla ricerca del tempo nell'arte*, Mondadori, Milan, 1954, p. 122.) The schematic man of art and of science do not coincide, but the painter expresses his images of man in his own way, just as science has firmly fixed man's biological attributes.

[1] F. Barone, *Il neopositivismo logico*, Edizione di Filosofia, Turin, 1953, p. 199.

case: rather they are prescriptions, rules of behaviour for the researcher in finding true propositions and predicting certain events."

Barone maintains that Schlick's position is not pragmatic, but on the problem of induction it appears to be eclectic since one feels in it the need for a logical justification, which is absent.

The point of view we have developed in these pages seems to be exempt from such a criticism, since it is frankly based on the consistent development of the classificatory procedure, that is, on the logical development of an experimental convention.

362   Russell on induction

Von Wright, after thoroughly discussing the theses of Hume and of other writers who, while affirming the *de facto* need of induction, agree, as does Russell, that it has until now remained unjustifiable, suspects that one is faced here with a confusion. "This confusion," he explains, "which is deeply rooted in the philosophical inclinations of man, and which is one of the fundamental sources of philosophy as such, consists in the failure to separate clearly questions of language from questions of fact. In almost any situation where there is alleged to be a conflict between 'philosophy' and 'common sense', the conflict can be shown to be the result of this confusion." In fact, the thesis on "the impossibility of justifying induction is — when rightly understood — grammatical in its nature and, as such, free from all implications of scepticism".[1]

The thesis of the present work is — if one wishes to see it in this light — both analogous and different. Here we maintain that induction would not be justifiable on the facts if by facts one meant the properties of the world extraneous to man, but that it becomes so when, in referring to events (that is, to sensory perceptions classified and expressed by common nouns), one postulates that these perceptions will continue, even outside present experience.

363   Forecasting the future as the norm of conduct

It is necessary to accustom oneself to the idea of a static future. This idea has been developed in these pages at some length so as to reconcile it with the undeniable fact that some events are very stable while others are very changeable, and hence not equally predictable. Physical phenomena would come into the first group, and social,

---

[1] G. H. von Wright, *op. cit.*, pp. 188 *et seq.*

political and economic phenomena into the second. We must conclude that accuracy in forecasting depends on the elements singled out for forecast; and that physical phenomena have been under examination since remote times and consequently their behaviour has been described in a very general and consequently regular form. The position regarding "social" events is different, since man is actively concerned in them, and the least details have an interest for him. Notwithstanding this, a large part of daily activity depends on the stability of such events.

Men gear themselves to the laws of economics and sociology. In spite of visible changes, they forecast future developments, they buy and sell, save money, assure a career to their children through education, and rely upon the stability of the political regime. In other branches of knowledge the forms and rules of language are considered as fixed, needs and tastes are postulated as unchangeable, and even in technical subjects — revolutionized these days — plant, machinery, and productive processes are standard. There is in all this, no doubt, a complex psychological component; but in so far as it concerns rational activities such as the forecasting of prices, of social and political dispositions, of productive systems and of means of communication in language, it is a useful (not to say necessary) hypothesis of work and of life, which produces all kinds of deductive development. If prices remain stable, saving money will enable future purchases to be made; if the political system does not alter, the social career of children will be subject to the same standards as that of their fathers; if the language is not revolutionized, a great poetical work will become immortal. It is not quite true that history is *magistra vitae*, because individual facts do not repeat themselves and hence they teach nothing; but it is true that the past, reduced to ordered categories, is a guide to men's conduct. The saving clause *rebus sic stantibus*, often invoked in the social sciences, is a kind of warning to conduct oneself so as not to attribute any absolute value to expectations which could be but transitory.

364 Rational links in induction

The inductive process which leads to the formation of common nouns is repeated when, beginning with these, the ordinary man and the scientist carry out more complicated tasks, bringing to light groups of similarities not immediately apparent, and sometimes forming mental constructs — "inductive–deductive structures" — in which common nouns are accepted as premises and in which it is difficult to disentangle the numerous factual and logical elements involved.

But the idea of the patterned mosaic, if it does not explain this process, at least contributes to an understanding of how it is possible. The subject has been closely examined, but perhaps some further explanation is necessary because of an opinion expressed by Professor Dingle, to whom this simile is due. "We must regard our puzzle-solver, then," he writes, "as having at his disposal two kinds of things. There are, first, the unit pieces of the puzzle — sense data — which he must accept as they are. He cannot alter them in any way, and his object is to connect them together into a rational unity. And, secondly, there are other pieces — clips, if you like — which he can use for joining the sense data together in a rational relationship. These he creates for himself and can alter them in any way he wishes so long as he does so consistently." Later he explains: "The clips are concepts."

It should be noted that the first of these series of pieces is, so to speak, fundamental, that is, it represents genuine natural entities only in so far as it is concerned with elementary sensations — the "elements" of Mach or historical data. But in passing to common nouns — that is, to linguistic expressions used in the constructs of science — the underlying objects, far from being "accepted as they are", have been adapted and transformed, as we have already discussed at some length. Then come the clips, the connective concepts, the rational links, which compose the mosaic whose patterns are called "pictures of the world" and "scientific doctrines". For the moment we are not concerned with going into the structure of these patterns; to do so would require many other ideas which have not yet been broached (Chapter 5); but the foregoing discussion already contains in embryo the entire process leading from historical data, from incomprehensible disconnected entities to their normalization in events — that is, to their coherent interpretation as needed in science and in everyday life.

365   Harrod's viewpoint

In section 346, a sentence of Harrod's was mentioned which it is as well to repeat: "If Newton's Laws of Motion were *grossly* wrong . . . it is singularly curious that we have found no motions inconsistent with them through all this time and over so wide a range of observation". Now the italicized adverb nullifies the whole point of this statement, because it is clear that the motions considered by Newton are those which he himself proposed and about which he theorized and which consequently obey his laws. The idea of "grossly wrong" is excluded from these laws. But everybody knows that when these concepts are more thoroughly examined, relativity theory has clearly defined the limits beyond which their validity lessens. By admitting that the so-called "cosmic constant" is equal to zero, or sufficiently small (note here also the approximation), the gravitational equations of Einstein are reduced to those of Newton.

But since there is no unique Newtonian potential which is applicable to all gravitational fields, in the same way there is no unique Riemannian space–time for all material systems. Therefore every physical situation corresponds in kind to a separate space–time. It follows that not even the most elaborate cosmology is true in the sense that it succeeds in explaining all the facts, nor that it is valid beyond doubt both for the past and the future. The question is always one of inductive structure that simplifies and interprets the world — a structure which is not summarized by a common noun and has need, on the contrary, of Newton's *Principia* and Einstein's scientific writings in order to be expounded, although it is of the same nature as the simplest classifications. Its future validity is just as hypothetical as the expectation that an acorn will grow into an oak tree and that a pendulum will continue to have isochronous oscillations.

366 Time in induction

Since we have repeatedly connected the idea of generalization with that of forecasting future events, it might seem superfluous to return to the time-dimension constantly used in the sciences.

In Italian, a word such as "time" has several meanings (section 131), provided that in each context its sense is clearly defined by convention. I refer not only to the different meanings of the word which other languages distinguish as "duration" (German *Zeit*), "weather conditions" (German *Wetter*), or "musical beat" (German *Tempo*)[1], but to those synonyms which could be confused. Historical time is one kind, which is concerned with entities and is unrepeatable and irreversible (sections 354 and 4347); another kind is the flux of consciousness so acutely analysed by Bergson, involving individual participation in events. Yet another is the reversible time about which theories are put forward in abstract sciences such as theoretical mechanics[2]. But the time in which events are immersed is different again: on the one hand it separates the past from the future, on the other it unites them in a

---

[1] For an analogy, see section 435.

[2] Precisely because mathematical time is different from physical time, Fantappiè could consider the mathematical time-dimension as reversible and could formulate his theory of "syntropic" phenomena in which time appears with a negative sign. Such phenomena, particularly biological ones, would be immune from entropic increase. L. Fantappiè, *Principi di una teoria unitaria del mondo fisico e biologico*, Rome, 1944.

classifying uniformity expressed by a common noun. Tasso's oak is past; the oak which the forester will mark for felling is, in a certain sense, future; but "the oak", common noun, is past and future together; that is, it is the expression of past experiences but interpreted as being indefinitely repeatable. Obviously, as a common noun is the complex result of voluntary acts, any alteration of its content and form will depend on new voluntary acts. In particular, the "time" of science could be regarded as a specific neologism, to distinguish it from other accepted meanings. Without wishing to go thus far, I found it necessary to have a clear understanding of the meaning of this word and its implications, which are bound up with the concept of induction developed here. In conclusion, owing to its regularity, or, better, to its predetermination, the scientific future can always be foreseen, and it is made more regular by the mathematization to which classified events are subjected.

The discussion here is still rather incomplete and merits a deeper analysis.

37   Recursive nature of the sequence hypothesis–deduction–induction

It is now convenient to revert to the ternary scheme of the scientific process: hypothesis, deduction and induction, as set out in section 322. The order of setting out and illustrating the three phases was arbitrary, and now that the traditional separation between them has been minimized, it seems even more so. Induction determines the hypothesis,

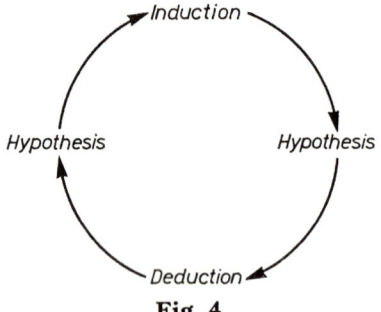

Fig. 4

and from the hypothesis, assumed as an axiom, the deduction is developed which, in its turn, suggests new hypotheses and necessitates a return to the facts from which still other inductions and hypotheses will emerge. The circularity of the process is clearly illustrated in the diagram, Fig. 4.

We shall return to this subject later.

38 Jeffreys' viewpoint

Our conclusion will certainly encounter scepticism, because it neither postulates the regularity of Nature, as did the ancients, nor affirms the immediate validity of inference, as upheld by Jeffreys; it does not even rely on the greater guarantee which short-term forecasts would give as opposed to long-term, as does Harrod.

Jeffreys' viewpoint is consistent with the practical self-reliance of scientists in their work — men who have not time for philosophical considerations and gain confidence from earlier successes. Jeffreys' book has an opening dialogue recalling the style of Plato and Galileo. To the astonishment of his interlocutor on hearing that science can make forecasts about "systems in which the number of possibilities is unknown and even infinite", the logician insists that "scientists are making this sort of inference all the time, without knowing much about how or why they do it. So I am sure that somebody should try. It would certainly need a lot of mathematics, and a good deal of knowledge of actual scientific procedure. The difficulty about expressing the ideas mathematically does not appear insuperable, considering how successful mathematics has been in extending its scope at various times."

Then, before developing the theory, Jeffreys briefly formulates a programme and asserts that "inference from past observations to future ones is not deductive," and that this process "is not only free from self-contradiction but is the only method possible".[1]

Enough has been said about Harrod's position, and a few words may sum it up. "The task before us is simple but arduous," he says. "We have to show that there can be valid inductions about Nature with no prior assumptions whatever about the characteristics of Nature. We must remain completely open-minded until the accumulation of evidence from experience begins to give guidance."[2]

Unfortunately the question is not so simple, and it is surprising that a man of subtle reasoning power should so frequently have recourse to common sense, to the likely and the reasonable, leaning rather towards psychology than logic (section 346). Out of respect for inferential rigour, one cannot easily approve of statements such as the following: "Suppose that Newton's Laws are destined to have a temporary reign only, and that the exceptions will be in the future; then it is not curious at all that we have not come across them. It does not

---

[1] H. Jeffreys, *Scientific Inference*, p. 12–13.
[2] R. F. Harrod, *op. cit.*, p. 23.

E*

follow that we must adopt an attitude of complete scepticism toward the future, although much current logical doctrine implies that we really should; to do so would grossly violate common sense. In this matter common sense is right; simple induction gives us valid grounds for attaching high probability to predictions, but only if these are sufficiently circumspect."[1] Hardly so; there is a high probability that Newton's laws will last for a further minimal fraction of time, but if the future instant hypothesized by Harrod at which they could cease to be valid was really this present instant, inductive forecasting, however short-term, would be just as wrong as that of the man who, on the basis of Fig. 2, after correctly forecasting the appearance of twenty-seven black balls, found himself at the point where the first of the three white balls were appearing; or as wrong as the other man who, taking a long view, considered the continued appearance of black balls valid for a millennium or more.

It was therefore right to maintain here, by logical reasoning, that the future, understood in the historical sense (section 366) and also in the inductive sense of other writers, cannot be conquered; whether experiments are many or few, or whether one looks far ahead or not, the insolubility of the logical problem remains the same. The whole chain of past entities and events can form a subject for reflection, but the non-existent chain of the future can only be reached by faith, that is to say, not by rational but by psychological methods. It is foreseeable only by hypothesis, that is, with a query to which only fact, and not logic, can answer yes or no[2].

These are psychological and subjective conclusions which recall the *a priori* non-demonstrable principles of Kant, of the type *lex parsimoniae* and suchlike (section 3325).

---

[1] R. F. Harrod, *op. cit.*, p. 73.

[2] One day towards the end of Wittgenstein's life the philosopher Norman Malcolm asked him if, "while he was writing the *Tractatus*, he had ever decided to choose an example of a 'simple object'; he replied that at that time he considered himself a logician, and it did not concern him, as a logician, to establish whether this or that thing was a simple object or a complex object, since the question was purely *empirical*. It was clear that he considered his former opinion absurd" (Norman Malcolm and Georg Henrik von Wright, *Ludwig Wittgenstein* (1958), Italian edition, 1960, Bompiani, Milan, p. 91. This episode, which confirms that at the time of the *Tractatus*, Wittgenstein considered facts extraneous to logic, agrees well with the thesis of these pages, namely that induction — the instrument of logic — cannot forecast even the most probable empirical data.

381 Repetitiveness, variability and the law of Gauss–Quetelet

In this section the connecting link is established between the above viewpoint and the ideas of previous authors, especially of Harrod, who maintains the validity of inductive forecasts until it is belied. For example, once we admit — although impossible to prove except by classificatory definition — that oak-trees will grow from acorns and that the synthesis of water will continue to occur under the same, most recently defined conditions, there follows as a practical certainty the development of those operational and cognitive activities on which the natural sciences have been built and to which they owe their triumphs.

It will later be shown that, in the past, the asserted regularity of classified events was contradicted by the so-called variability of natural phenomena, and that not until the first decades of the nineteenth century did the mathematicians Gauss and Quetelet (the second following on from the first) succeed in reconciling the two antagonistic viewpoints and thus, with an appropriate theory, opening the field to immense developments. In other words, it was necessary to conceptualize the experimental data by mathematical means in order to bring back variability into the ambit of regularity. This notion accords with the thesis maintained in this study, which has its linguistic core in the common noun. Mathematics in science does not express an objective status of Nature, as was believed by Galileo and perhaps by Quetelet himself, but is only a useful "clip" allowing classified events to be connected.

All objects known by the same name are different, and yet the same; when repeatable, they can be distinguished individually by details. But division into classes and the related verbal descriptions eliminate the details, reducing them to irregularities, and so creating a regularity which is itself the classificatory scheme. When such objects are measured for this or that distinguishing characteristic, they certainly reveal differences, but these never exceed the limits beyond which an entity would cease to be itself and would become something else — as though, for example, a small camomile flower (*Matricaria chamomilla*) could become a giant sunflower (*Helianthus annuus*).

But this too is an artificial example. Suppose for a moment that botanists were to consider the sessile florets of a flower as the only point of resemblance common to certain herbaceous plants. In such a case, all the Compositae, from marguerites and thistles to sunflowers, would be included in one class rather than distributed among the numerous "species" of that "family" of phanerogams. The seeds of the plants of this supposedly single species would at times produce a camomile and at times a sunflower, apparently at random. The variability would be great, but would almost certainly conform to the mathematical law of Gauss–Quetelet. By restricting the classes, botanists have greatly lessened the variability and enabled more precise forecasts to be made[1].

382 Variability and arbitrariness in classification

The arbitrariness of this procedure, which does not correspond to the Linnaean idea of *Systema Naturae*, will be stressed by considering classifications based on measurements. In section 353 mention was made of the division into five classes of the cephalic indices referred to by Hempel. Below, the quantitative limits used by him are compared with those used by a respected Italian anthropologist, R. Biasutti[2]. (The nomenclature has been brought into line.)

Name of cephalic form	Relative cephalic width per 100 of length	
	Hempel	Biasutti
Dolichocephalic	75·0	74·9
Sub-dolichocephalic	75·0–77·6	75·0–78·9
Mesocephalic	77·6–80·0	79·0–82·9
Sub-brachycephalic	80·0–83·0	83·0–86·9
Brachycephalic	83·0	87·0

It can be seen that, because of the arbitrary fixing of limits, some numbers come into different groups. For example, the brachycephalic

---

[1] The point is emphasized when we note that classes become more restricted as the number of classified characteristics increases. Years ago there was an assistant in an Italian University who devoted his whole life to compiling, unsuccessfully, a list of aquatic living species. An enormous task, to group under this one characteristic bacteria, algae, infusoria, amphibians and fish, aquatic birds, hippopotami, and cetaceans. Clearly the undertaking would have been halved merely by limiting the research to animals only. Similarly, anyone wishing to list all bloodsucking animals would include in a single group fleas, leeches, vampires, etc.; but the sanguineous mammals would be few, and sanguineous men would be reduced perhaps to the Masai. Classes would become ever more restricted, with ever less variable attributes.

[2] R. Biasutti, *Le razze e i popoli della terra*, vol. 1, Utet, Turin, 1953, p. 238.

for Hempel are the sub-brachycephalic for Biasutti. By means of classification, therefore, many phenomena can appear repeatable or at least more regular than they would be in a different framework.

Scientists understand this and avoid modifying classifications, especially in the more complex situations where unnecessary confusion could result. We have said that the synthesis of water is still thought of today as it was in the last century (section 3555); and Newton's laws are still upheld, notwithstanding the generalization of relativity theory. And again, the incessant subatomic movement is enough to make every object different from itself; but this can be left out of account in defining objects, so that, for instance, the writer and the reader of these pages can both remain undisturbed at their respective work-tables (cf. section 11).

39 Classificatory induction and the strengthening of science

Science is not weakened but strengthened in the light of these considerations. It matters little whether the world is ruled by laws as understood by ordinary men; but it does matter that the human mind, bringing order into the chaos of external information, should be able to formulate it with the three-fold purpose of satisfying the ever-recurring desire for knowledge, of harnessing Nature to the needs of men, and of communicating man's achievements. It is hard to see why it should be a more noble aim for thinkers to discover the secret designs of Nature than to build them up by creative imagination from factual data. The pattern of science, constructed and adapted from concepts of Nature, is more harmonious, more stimulating, more useful than objectivity as understood by the Positivists and their followers until the beginning of the present century. Similarly, a clever chess-player is prouder of the imaginative position which he can plan, than of discovering by what mysterious device an automaton could checkmate him in seven moves.

It should, then, be admitted without hesitation that experience in itself is changeable and unpredictable: but events classified and expressed linguistically take on, so to speak, a timelessness which allows them to be considered as fixed and repeatable until fresh knowledge and ideas require them to be modified. The problem of induction and of the future is thus reduced to human dimensions, or rather to one of logical form, which validates deductive inferences. Obviously, since we are concerned with the adaptation of the world to man, and not with the interpretation of the world from outside, we do not presume to penetrate the mysteries of the universe by means of the theory of induction.

To seek the essence of things, wrote Galileo, is an impossible undertaking and a vain task; but, before him, Dante (*Paradiso*, xxiv, 64–5), paraphrasing St Paul (Hebr. 11, 1), had written — according to the commentary by Ottimo[1] — that only by faith and not by human knowledge can man come to understand that "*substance* is a thing certainly known and evidence leading to proof" of things not seen.

(Hebr. 11, 1) "Now faith is the substance of things hoped for, the evidence of things not seen."

---

[1] The quotation from Ottimo is from the commentary on the *Divina Commedia* by N. Sapegno, Ricciardi, Milan–Naples, 1957, p. 1076.

4

Probability and statistics

41   The vogue of gaming in the seventeenth century; origin of the theory of probability

Sir Ronald Fisher has written that the formation of the concept of mathematical probability — unknown to the Greeks and to the Islamic mathematicians of the Middle Ages — was perhaps made possible "by the great favour in which the French and English nobility of the seventeenth century held the recreation of gaming, and by the advanced techniques of the time in the preparation of the necessary implements, such as dice, cards, etc., with sufficient precision to allow of mathematical calculation".[1]

As will be seen later, this is truer than might at first appear. When Greeks and Romans laid bets on the faces of the huckle-bones of sheep, it cannot have been easy to determine any reliable rule for the outcome of the throws. A Renaissance writer appears to confirm this in mentioning the number of points allotted to the various throws, and he gives the impression that it was done in a quite arbitrary way. Professor F. N. David — who discusses the question in a scholarly study, one of a series on the History of Probability and Statistics edited by Dr M. G. Kendall[2] — has carried out a series of throws with small bones and with Greek and Roman dice recovered from archaeological excavations,

---

[1] Sir Ronald Fisher, "Mathematical Probability in the Natural Sciences", in *XVIII International Congress of Pharmaceutical Sciences*, Brussels, 1958.

[2] F. N. David, "Dicing and Gaming (a Note on the History of Probability)" in the series *Studies in the History of Probability and Statistics*, No. 1, *Biometrika*, vol. 42, parts 1 and 2, 1955 [reprinted in *Studies in the History of Statistics and Probability*, ed. E. S. Pearson and M. G. Kendall, Griffin, London, 1970]. See also her *Games, Gods and Gambling*, Griffin, London, 1962.

and has demonstrated the considerable inequalities of the outcomes, due to their imperfections.

But apart from the history of the instruments themselves, interest in the results of games of chance really arose in the seventeenth century when some of the most authoritative names became involved, such as Cardano, Galileo, Pascal, Fermat, Huyghens, and others. This shows that the interest was linked not so much with a passion for gaming as with the spirit of modern science and with the fact, noted and stressed by Pasteur, that problems become evident and find solutions only with receptive minds (section 54).

411 Probability and Statistics

We are, however, concerned not with the history but with the concepts of probability and later of Statistics — two words not yet in the metalanguage of this book, because they would have seemed incomprehensible verbal forms without the premises of the previous chapters. They need now to be introduced with a thorough explanation.

412 Diderot's concept of probability

Diderot, working within the framework of Cartesian and Leibnizian rationalism and extending ideas already put forward by James Bernoulli, wrote as follows in an excellent long article in the *Encyclopédie*: "Nous pouvons apercevoir plus ou moins les relations qui peuvent être entre deux idées, ou la convenance de l'une avec l'autre, fondée sous certaines conditions qui les lient, et qui, lorsqu'elles nous sont toutes connues, nous donnent la certitude de cette vérité, ou de cette proposition; mais si nous n'en connoissons qu'une partie, nous n'avons alors qu'une simple *probabilité*, qui a d'autant plus de vraisemblance que nous sommes assurés d'un plus grand nombre de ces conditions. Ce sont elles qui forment le degré de probabilité. . . ."

Further on, explaining that, of these two connected ideas, one might be a certain sign (*signe*) of an uncertain thing, he adds: "The barometer falls, it is a sign of rain; the sign is certain but the rain is doubtful, because the barometer often falls without rain following." He continues: "It follows that the conclusion drawn from a cause or from a sign whose existence is certain has the same degree of probability as the outcome of that cause or sign."

It would be difficult to state a clearer or more correct idea of probability in its current meaning. Diderot develops it at length, and having also introduced, in the passage referred to, the notion of degree of probability, finds that he has opened the way to the mathematical notion[1].

413 The logical and the mathematical concepts of probability

One therefore sees — as Poincaré warned — that the problems of probability present two distinct aspects: the first, so to speak, "metaphysical" (today it would be called "logical" or "epistemological"), which legitimizes the various conventions, and the mathematical, which applies the conventions themselves to the rules of calculus[2].

Many authors, from Boole to Keynes, Jeffreys and Carnap, have reconsidered the Leibnizian aspiration to a logic of probabilities,[3] refining the basic concepts, and have, so to speak, superposed a new structure on the achievements of earlier mathematicians — not infrequently rejecting them[4].

We shall try here to overcome the difficulties of a historical account and examine separately both points of view, looking for a synthesis which can reconcile them.

---

[1] In ordinary parlance the synonyms "probable" and "likely" are used interchangeably, but always in the sense of events which happen fairly frequently. Perhaps the meaning of other ordinary words such as accidental, fortuitous, chancy, and so forth (some of which occur even in scientific language) has shifted in the opposite direction. Apropos of this, it is worth noting the following well-contrived sequence from a French dictionary of synonyms: "Stones fall *accidentally* from the sky; an idea comes *fortuitously* to mind, and it is pure *chance* if a stupid person has a good idea." (H. Bénac, *Dictionnaire des synonymes*, Hachette, Paris.)

[2] H. Poincaré, *Calcul des probabilités*, Gauthier-Villars, 1912, p. 29.

[3] In an epigraph to the first chapter of his *Treatise on Probability* Keynes quotes the following from Leibniz: "J'ai dit plus d'une fois qu'il faudrait une nouvelle espèce de logique qui traiteroit des degrés de Probabilité" (Keynes, *A Treatise on Probability*, Macmillan, London, 1929, reprinted 1952). For the other authors quoted see G. Boole, *The Laws of Thought* (1854), Dover Publications, New York; H. Jeffreys, *Scientific Inference*, Cambridge University Press, 2nd edn, 1957; R. Carnap, *Logical Foundation of Probability*, 1950 (2nd edn, 1962). Only the first part of Carnap's great work has been published; there is an abridgement called *The Continuum of Inductive Methods*, University of Chicago Press, 1952.

[4] The opposing tendencies result from the long critical review of the work of Keynes by E. Borel which appeared in the appendix to the volume "Valeur pratique et philosophie des probabilités", in *Traité du calcul des probabilités et de ses applications*, vol. IV, fasc. 3, Gauthier-Villars, Paris, 1939, p. 134.

42   Five lines of development in the theory of probability

Carnap[1] gives a classification of the various concepts of probability which, conveniently integrated and brought up to date, enables five different lines of development to be listed:

(a) There is first of all the classical conception originated by James Bernoulli, systematically developed by Laplace, and presented by their successors in various forms; here probability is defined as the ratio of the number of favourable cases of an event[2] to the number of all possible cases, provided they are equally possible. This is a definition axiomatically assumed for analytical purposes and therefore unexceptionable in itself.

(b) The conception of probability as a kind of objective relation between propositions (or sentences); the chief exponents of this concept, besides Boole, are Keynes and Harrod.

(c) The conception of probability as the relative frequency of occurrence of a particular modality of an event with respect to the occurrence of all the modalities. Formulated many years ago in Italy by G. Mortara, it was later transformed, made axiomatic and fully developed by von Mises, Kolmogorov, Reichenbach and Wald, and is accepted in different ways by the majority of mathematical statisticians (Fisher, Neyman, E. S. Pearson).

(d) In contrast with the two preceding positions can be placed those writers for whom probability would be the degree of belief in the occurrence of an uncertain event. The idea of a "subjective" probability is not recent, but only since the 'thirties has it been systematically developed with inductive aims, initiated by de Finetti and followed by L. J. Savage, perhaps by Anscombe and by others. In the theory of Jeffreys, too, can be discerned points of contact with the aims of the subjectivists.

(e) Finally there is the position of Carnap, who distinguishes clearly between different probabilities, both regarded as inductive categories. He speaks of a "probability$_1$", to be understood as "degree

---

[1] R. Carnap, *op. cit.*, pp. 24–5.

[2] In this notion of probability (section 42 *et seq.*) the word "event" is used in its proper sense (section 354), that is, relative to impressions, or complexes of classified impressions, which are repeatable and are denoted by common nouns. "Events" in the sense of the probabilists are: a court card turning up when a card is drawn from a pack, the appearance of an odd or even number in a lottery, the ball or needle stopping on red or black in roulette, the birth of a son as opposed to a daughter, the finding of a four-leaved clover in a field of clover, and so on. Various examples of this type will recur later.

of confirmation" of a hypothesis, which would be near to but not identical with the position of the mathematicians; and of a "probability$_2$", to be understood as "relative frequency", which would be near to the position of the statisticians.

In a chapter of a work such as this, having a different purpose, it would not be possible to go thoroughly into the theories of all the writers quoted. Here only a few significant viewpoints will be examined, with special reference to the problems most nearly concerning the material of this book[1].

421　The mathematical notion and its initial problems

Since the theory of probability was the outcome of calculations based on games of chance, obviously the first, most numerous and authoritative people to study it were mathematicians. Only later was a logic of probability and its principles developed.

The notion (already referred to) of probability in a game of chance as the ratio of the number of favourable cases to the number of possible cases of an event, assuming them to be all equally possible, had already been correctly stated by Cardano (*circa* 1520) in his book *De ludo aleae*[2]. It came under scientific scrutiny by Pascal and Fermat, and is so well known as to need no special mention, unless for the sake of balanced exposition. According to this concept, when a coin with two faces is tossed, it can come down heads or tails, and the probability of either side appearing is $\frac{1}{2}$. Similarly, the probability of one particular face of a die coming up is $\frac{1}{6}$; the probability of drawing a certain number in tombola (a kind of lottery) is $\frac{1}{90}$; the probability that the first draw in a lottery is odd or even is $\frac{1}{2}$; the probability of drawing the seven of hearts from a pack of cards is $\frac{1}{52}$, but is $\frac{4}{52}$ for a seven of any suit. The probability of the number backed at roulette coming up is $\frac{1}{36}$ (not including zero), while the odds are $\frac{1}{2}$ for *rouge* or *noir*.

It is agreed that all these numbers, and many others which could serve as examples, measure probabilities because they are ratios be-

---

[1] A rapid review of some modern theories on probability may be found in the volume: G. H. von Wright, *The Logical Problem of Induction*, Blackwell, Oxford, 1957, pp. 90 *et seq*.

[2] C. Gini, "Girolamo Cardano e i fondamenti del Calcolo delle Probabilità", in *Metron*, vol. 19, No. 1–2, 1958. M. Boldrini, *Statistica, teoria e metodi*, Giuffrè, Milan, 1962, p. 36. Luca Pacioli is also mentioned. Compare S. Wilks, preface to *The Book on Games of Chance*, Holt, Rinehart & Wiston, New York, 1960. In recent years many other books have been written on Cardano, and his works have been published in English. See F. N. David, *Games, Gods and Gambling*, p. 54.

tween cases favourable to an event and the total number of cases admitted to be *a priori* equally possible. Note the inherent limitation in the condition of apriority, which means that one speaks not of experimental results but of abstract numerical relations with an axiomatic meaning. Probabilities are always expressed by proper fractions because they state the ratio of parts to the whole (for example, heads, with a coin of two faces; odd numbers out of all the faces of a die, and so on). From the relevant axioms, mathematicians have evolved all the theorems of the Calculus of Probability by deduction. And it is precisely in the development of the theory that difficulties appear from time to time, connected either with the computation of chances or with the guarantee of equal possibility and independence.

It was this kind of difficulty which led the gamesters of the seventeenth century to turn to Galileo and Pascal for the solution of gaming problems that seemed incomprehensible puzzles. There was a game called "passes ten", in which three dice were thrown[1] and equal bets made on a number turning up greater than ten, or ten and less. Galileo was asked why the numbers 9, 10, 11 and 12 turned up with different frequencies, despite the fact that they could all be obtained by six

---

[1] The game with three dice, studied by Galileo in the minor work *Sopra le scoperte dei dadi* (referred to in vol. V of the *Works* in the Salani edition, Florence, 1935, p. 387), is very similar to the game of "zara" mentioned by Dante in the opening scene of the sixth canto of the *Purgatorio*. In the commentary by Sapegno on the *Divina Commedia*, it is explained that the game was widely played in the fourteenth century, that it had been forbidden, to no purpose, by many municipal statutes, and that it was played by throwing three dice on to a table. "Those numbers", adds the commentator, "below 7 or above 14, such as 3, 4, 17 and 18, which could occur by only one combination rather than several, were considered null; 'therefore, when those numbers turn up, the players say "zara", as if to say "nothing", like zero in the abacus' (Buti)". Clearly it is true that those numbers are all composed of a single partition, but while 3 and 18 can occur in one way (as 1, 1, 1 or 6, 6, 6), 4 and 17, on the other hand, can occur in three different ways (as 2, 1, 1; 1, 2, 1; 1, 1, 2 etc.). It is the same misunderstanding which Galileo perceived and corrected in the sixteenth century. He expressly pointed out "the 3 and 18 as numbers which can be composed in only one way with three numbers". And further on he adds: "The number which is composed from the three numbers, two of which may be the same and the third different, can be produced from three permutations: as for example 4, which can come from 2 and two aces, can occur in three different ways. . . ."

[*Translator's Note* — Sapegno's derivation of "zara", the old form of hazard, is in error. The game was probably brought back to Europe by the Crusaders and derives from the Arabic *al zhar*, meaning "the die". The appearance of the initial h in "hazard" is an etymological mystery which has never been cleared up.]

combinations of numbers. Galileo showed that this reasoning was incorrect and that, while 10 and 11 could be formed in 27 different ways, 9 and 12 could be formed in only 25 ways, so that, in the long run, these last two turn up less often than the other two. It is easy to see that, in throwing three dice a, b and c, there are 216 ($6 \times 6 \times 6$) possible cases. But it is more difficult to count the number of cases favourable to various totals, as for example those adding up to 9. The possible combinations are the six listed in the following table, as correctly judged by the players who questioned Galileo.

1st combination	1	2	6	happens in 6 ways
2nd „	1	3	5	„ „ 6 „
3rd „	2	3	4	„ „ 6 „
4th „	1	4	4	„ „ 3 „
5th „	2	2	5	„ „ 3 „
6th „	3	3	3	„ „ 1 way

The first three "combinations", however, can occur in six different ways because each of the numbers can come up on die a, b or c. On the other hand, the fourth and fifth combinations can happen in only three ways, given that the odd number may turn up on die a, b or c and the even numbers on the two remaining dice. Finally, the sixth combination is clearly unique. Hence the 25 possibilities enumerated by Galileo. The probability of 9 turning up with three dice is therefore 25/216 and not 6/216.

422 An unsolved problem

But take an example still difficult to solve. Given two similar packs of playing cards, suppose that one card is drawn from each pack and placed face downwards on the table. One card is taken up and found to be black. What is the probability that the other is black also?

Poisson (1837) reasoned as follows. Before the card is turned up there are three equal possibilities, black–black, black–red, red–red. The first card is turned up and is shown to be black. This means that the alternative, red–red, is excluded. If both cards are black, we are twice as likely to have turned up a black card as if only one is black. There are thus two chances favourable to a second black card, only one to red, and the probability that the second card is also black is therefore $\frac{2}{3}$.

Years later, von Kries objected that the card drawn from the second pack has nothing to do with that drawn from the first; and since in the second pack there are just as many black as red, the probability that the second card is also black is $\frac{1}{2}$.

The difference between the two answers derives from differences as to what cases we admit as initially equally probable. There being an element of choice in the assumptions on this point, Poincaré—in agreement with Keynes, from whom the discussion is taken—admitted that until some more convincing basis of the problem is found, the choice between the two solutions is arbitrary.

423   Laplace's condition of equal possibility

I would not have dwelt so much on these examples had they not seemed necessary to underline the need for exact calculation and, in particular, for a clear specification of what possibilities are considered to be initially equally probable.

In his first researches, Pascal had already commented on this. (It is intuitively felt that, if the dice are lop-sided, or the cards marked, or the balls in the lottery of unequal size and weight, the chances are unpredictably altered.) It was James Bernoulli who systematically developed the concept of equal probability; and Diderot, who was inspired by him, notes with his usual clarity and penetration: "When we have ascertained that a certain event can happen in a determined number of ways, and we know or suppose that all these ways are equally possible, we can say with certainty that the probability of its happening in a given way is worth so much or is equal to so many 'parts' of certainty."

It was Laplace who drew the attention of the moderns to the importance of equal prior possibility. In the first pages of the *Essai philosophique sur les probabilités* he says: "We know that of three or more events one only can happen, while nothing leads us to believe that one of them will happen rather than the others. From this state of indecision it is impossible to say with certainty what will happen. . . . The theory of chances consists of reducing all events of the same kind to a number of cases which are equally possible, that is, are such that we are equally undecided about their realization. . . ."

In some circumstances, the randomness—and hence the equal possibility of the chances—can be satisfactorily stated. If we take, for example, the last figure of each logarithm in a table with many decimal places and count the zeros, ones, twos, etc., we find that each figure appears roughly the same number of times. Clearly it is as if the ten Arabic integers had been drawn out of a bag and replaced from time to time. If, on the other hand, the results turned out to be notably different (and the adverb can be measured and estimated probabilistically), we should have to conclude that something had hindered certain results in favour of others[1].

With his usual caution, Laplace added to the passage quoted above a warning concerning the difficulties presented by the reduction to equiprobable cases: "The theory of probabilities involves such fine points that it is not surprising if two people, with the same data, arrive at different results, especially in very complicated questions". The example of the two cards (section 422) shows the timeliness of this warning.

424 From Laplace to Poincaré

The greatest French probabilists, starting with Cournot and Poisson, were disciples of Laplace; and a hundred years later his ideas were still being held unchanged by men of the stature of Poincaré, Borel and Darmois.

Poincaré, for example, states that in a game such as roulette, the probability that the needle will stop at this or that number is a continuous and uniform function of the impulse, and hence, given that a small variation of the impulse to the rotating needle is enough to stop it at a different sector, the probability of any particular red sector coming up is the same as that for the following black sector, and the total probability for red is identical with the total probability for black[2].

425 Summation and product of probabilities; Bayes' theorem

Once the concept of mathematical probability was fixed, the classical exponents of the calculus moved on to a series of developments, indicated by the following simple but basic theorems.

(a) *Problem of summation.* Given two events E_1, E_2, which are mutually exclusive (that is, only one or the other can happen), with

---

[1] H. Poincaré, *op. cit.*, pp. 22 and 313.

[*Translator's Note* — Remarkably enough, this process does not give a sequence of random digits. Cf. J. Franel, *Vierteljahrschrift der Naturforschenden Gesellschaft in Zürich* (1917), vol. 62, p. 286. For a discussion of the problem of ensuring equal prior probabilities in sampling, see M. G. Kendall and A. Stuart, *The Advanced Theory of Statistics*, Griffin, London, vol. 1, 3rd edn, 1969, Chapter 9.]

[2] H. Poincaré, *op. cit.*, p. 12. This discussion is developed, together with extracts from French probabilists, in M. Boldrini, "La théorie des probabilités", Discours Présidentiel à la 33ème Session de l'Institut International de Statistique, in *La Scuola in Azione*, 1961–2, No. 1; and also in *Revue de l'Institut International de Statistique*, 1961, No. 1.

respective probabilities p_1 and p_2, the probability that one of them happens is equal to $p_1 + p_2$. It can immediately be seen that, if in a bag there are respectively a, b, c balls of three colours, m in all, the chance of drawing one of the first two colours depends on the total of cases favourable to them, $a + b$, divided by the total m.

(b) *Problem of the product.* Given that two independent events E_1, E_2 are compatible (that is, that they can occur together), with respective probabilities p_1 and p_2, the probability that they both occur is equal to $p_1 p_2$. It is clear that if one of two bags contains a white balls in m and the other b white balls in n, each white ball from the first can be paired with each white ball from the second, and there will result a number of favourable combinations of two white balls equal to ab. Analogously, the number of pairs which can be formed with all the balls of both is mn, and the probability of drawing two white balls, one from each bag, is $(ab)/(mn)$.

(c) *Bayes' theorem* (either of the "probability of causes" or of the "inverse probability"), demonstrated at the end of the eighteenth century by the English mathematician of that name, has become the apple of discord among probabilists, logicians, and statisticians. Here is its demonstration as formulated by Bertrand[1].

Suppose that one of an exhaustive and mutually exclusive set of causes, E_1, E_2, . . ., E_n, has produced an observed event. Let the probability of these causes before the event happened be ω_1, ω_2, . . ., ω_n. The event takes place: the cause E_i gives the probability p_i to the event. The question is, what is the probability of each of the n causes which could have generated the event? The solution is very simple. Consider an urn containing both black and white balls, each one of which bears one of the numbers 1, 2, . . ., n. We can regard the occurrence of a number as corresponding to the cause with that number, and the relative proportions of the colours of the balls as determining the probabilities p_i.

What then is the probability that the ball drawn out will have a given number?

Let μ be the total number of balls; μ_i that of the balls marked with an i; and among these μ_i let the number of white balls be m_i. The total number of white balls will then be

$$m_1 + m_2 + \ldots + m_i + \ldots + m_n.$$

They are all equally possible, and the probability of the white ball drawn out being marked with i is

---

[1] J. Bertrand, *Calcul des probabilités*, Gauthier–Villars, Paris, 1907, p. 139.

$$P = \frac{m_i}{m_1 + m_2 + \ldots + m_n} \tag{1}$$

Also, the probability that a white ball marked with i is drawn from the set generated by the ith cause is

$$p_i = m_i/\mu_i, \quad \text{from which } m_i = p_i \, \mu_i.$$

Putting $\mu_i/\mu = \omega_i$, we have $m_i = p_i \, \mu \, \omega_i$.

Substituting in formula (1) for m_i and eliminating μ, we have

$$P = \frac{p_i \, \omega_i}{p_1 \omega_1 + p_2 \omega_2 + \ldots + p_n \omega_n}, \tag{2}$$

expressing the probability that the event was produced by cause i.

43   The logical notion of probability: Keynes's position

Writers concerned with logic have put forward searching criticisms of the linear definition of probability and of the simple formulation of the basic principles of the probability calculus given by mathematicians. The zeal of these critics for thorough inquiry has especial importance for the main subject of this book.

Keynes[1] avoids the temptation of formally defining probability, limiting himself to illustrating the concept. The terms "certain" and "probable", he states, express the various degrees of rational belief in a proposition, based on the knowledge one possesses. All propositions are true or false; hence the idea of probability is always linked with a nucleus of knowledge, and is not an objective characteristic of the propositions themselves.

In this sense, Keynes adds, probability can be called subjective, but this does not mean that it is at the mercy of caprice. It is not related to what one believes, but to the objective knowledge which one possesses; hence the logical character — that is, rational and cognitive rather than psychological — of the doctrine concerning it. Given a person's complex of knowledge, assumed as definite premises, the theory indicates what further consequences, certain or probable, can be drawn from it.

To these useful clarifications Keynes adds another, which also coincides perfectly with the point of view of this book. He rejects the use of the word "event" (see note to section 42), which recurs in all writers on probability, affirming that "it will be more than a verbal improvement to discuss the truth and the probability of *propositions*, instead of the occurrence and the probability of *events*". Apart from

---

[1] J. M. Keynes, *A Treatise on Probability*. See especially Chapter IV.

the harmless avoidance of a word — to which too much weight should not be attached — one should notice the "linguistic" formulation of the idea of probability, accentuating its logical content. "Between two sets of propositions, therefore", Keynes explains further on, "there exists a relation in virtue of which, if we know the first, we can attach to the latter some degree of rational belief. This relation is the subject-matter of the logic of probability." With no radical difference, Diderot had spoken of "perceiving the relations which can exist between two ideas". But since ideas can only be communicated by words, Diderot, without expressly saying so, also had to concern himself with pro-positional relations.

431 Jeffreys' position

Jeffreys' position does not differ greatly from that of Keynes. He avoids giving a formal definition of probability and, like the latter, proceeds without spending much time on the considerable problems of definition. "There is no harm", he declares, "in saying that a pro-bability expresses a degree of reasonable belief" or, as Carnap says, "the degree of confirmation of a hypothesis".[1] Let us call a such a degree of confirmation of the proposition (or hypothesis) q, given p. That is, $P(q|p) = a$; and it is not necessary that a should have a numerical value, though it is often convenient to fix numerical rules of transformation of this and of the other propositional symbols.

432 Harrod's position

According to Harrod, on the other hand, an explicit definition of probability is both possible and desirable. In stating this, he takes up this clear position: "Some theorists have held that probability is in-definable. I reject their view."[2] And he goes on to affirm that probability is the "relation between a premiss or premisses and a conclusion", provided that by premiss is understood evidential value. He therefore also agrees with Diderot and basically with Keynes himself. But it is in the logical development of the premisses that the two English writers part company.

From Harrod's definition it is clear that there can be propositions of certainty — such as "B always follows A" — and propositions of probability — such as "B sometimes follows A",[3] and consequently

---

[1] H. Jeffreys, *Scientific Inference*, pp. 22 *et seq.*
[2] R. F. Harrod, *op. cit.*, p. 48.
[3] Compare Diderot: "The barometer often falls without rain following".

propositions of probability admit of degrees. The way in which Harrod sets out the problem of induction has already been fully discussed in Chapter 3, and the argument will not be repeated here. It suffices to mention Harrod's formula of unconditional simple induction (section 345), namely the relation $x/(x+1)$, and point out how and why it can be considered as a measure of probability.

A person who — in harmony with Locke's conception — is completely ignorant of what experience may reveal to him travels along a continuity (such as crossing the Sahara, or moving along an endless street, or the running through of a typewriter ribbon), and remembering past experience, he foresees the results of experience to come. Suppose that he believes in the continuance of the continuity for at least $1/x$ of the length for which it has already proceeded. If he entertains this belief from the beginning of the journey to the end, he will be right x times for every once that he is wrong. This is in accordance with the traditional notation which gives a probability of being correct of $x/(x+1)$. The probability of the belief in continuance will be higher, the more modest the extrapolation. The relation $x/(x+1)$ (or whatever quantity one obtains by varying the length of the forecast) is therefore always a proper fraction and is likewise a measure of probability, that is to say the probability of carrying out an exact induction. In fact, it expresses the uncertain relation between the premiss A (previous experience) and the consequence (expectation) B by means of a proper fraction which can be interpreted in conformity with mathematical probability.

Harrod gives the best of motives for wishing to express induction and the corresponding probabilities in a generic form rather than in the traditional context of measurement, but he lacks clarity. It was mainly in order to bring the formula of induction into line with the customary mode of expression of mathematical probability that we found it preferable to introduce, in sections 342 and 343, examples in which the measures of induction could be expressed by enumerable discontinuous quantities such as clicks of a typewriter, or black and white balls which run in pre-arranged rings.

In any case, the fact most worthy of note is that induction, and hence probability as defined and measured by Harrod, are completely free (or are perhaps if we exclude the appeal to memory) from every apriorism and every alogical element.

4321 Logical formulation of summation and product of probabilities, and of Bayes' theorem

Modern authors sometimes develop probability in a generalized form — as opposed to the purely algebraic one adopted by the classicists (compare the example in section 425) — and in such a way as to extend its scope from quantities to propositions using logical symbols, the most basic of which have been indicated in section 163. This generalization was needed to facilitate the passage from purely propositional expressions (qualitative) to numerical ones (quantitative). Boole had already done this (1854), as also had Keynes, Carnap, Jeffreys, Cox, Wright, Savage and many others[1]. Instead of probabilities expressed as numerical relations, generalization makes it possible to speak of probabilities of propositions — probabilities which, by suitable devices, can be made quantitative and variable between 0 and 1. Included among them are those expressing the results of games of chance. For comparison, there follow here in a modern form the same three theorems as were considered in section 425, in the formulation supplied by Jeffreys[2].

(a) *Problem of summation* (already mentioned). From now on we let q, r, . . . be propositions and let p be the hypothesis or initial body of knowledge from which the propositions are obtained. The probabilities P of the propositions q, r, . . ., given p, are expressible quantitatively by values from 0 to 1, and the concepts of equality, summation, product, transitivity and so on are considered applicable.

As in section 431, the probability or degree of confirmation of the proposition q, given the hypothesis p, is denoted by $P(q|p) = a$. Since the purely propositional formulation does not have a quantitative character, Jeffreys, in order to make possible the change to quantities, requires the following three postulates:

(1) On data p, if q and r are two propositions, the first is more probable, equally probable, or less probable than the second, and the three alternatives are mutually exclusive.

(2) The relation of different or equal probability is transitive, so that if $P(q|p) \gtreqless P(r|p)$ and also $P(r|p) \gtreqless P(s|p)$, then $P(q|p) \gtreqless P(s|p)$.

(3) On data p, if q and r are mutually exclusive, and q' and r' are also mutually exclusive, and if q and q' are equally confirmed by p, and

---

[1] All the authors mentioned in the text have already been quoted, except H. T. Cox, *The Algebra of Probable Inference*, The Johns Hopkins Press, Baltimore, 1961.

[2] H. Jeffreys, *op. cit.*, pp. 24 *et seq.* Boole accepts the first principles of probability from the ideas of the classicists and refers to the solutions of the first theorems given by them (p. 249). But later (p. 257) he introduces logical calculus to formulate a more general theory, which is dealt with and developed by H. T. Cox.

r and r' are equally confirmed by p, then $q \lor r$ and $q' \lor r'$ are equally confirmed by p.

From this it is seen that, given p, when q and r are mutually exclusive, the numbers which correspond to the degrees of confirmation of q, r and $q \lor r$ satisfy the rule:

$$P(q \lor r|p) = P(q|p) + P(r|p),$$

which resolves the problem of summation.

(b) *Problem of the product.* It is necessary to add a fourth postulate to the preceding ones. Given the hypothesis p, if q and r are mutually compatible propositions, they cannot together be confirmed more strongly than each would be separately. In fact, q is equivalent to $(q \cdot r) \lor (q \cdot \sim r)$, while q, r and $q \sim r$ are mutually exclusive; hence the respective probabilities are summable.

From this it follows that:

$$P(q|p) = P(q \cdot r|p) + P(q \cdot \sim r|p) \leqslant P(q \cdot r|p).$$

If the probabilities of q and r are both zero, $P(q \cdot r|p)$ is also zero.

Furthermore, if both q and r follow from p, then $q \cdot r$ follows from p, and

$$P(q|p) = P(r|p) = P(q \cdot r|p) = 1.$$

From all this there is derived:

$$P(q \cdot r|p) = P(q|p) P(r|p),$$

which is the product rule.

However, to avoid antinomies which are present in particular cases. we generalize the product rule as follows:

$$P(q \cdot r|p) = P(q|p) P(r|q \cdot p).$$

(c) *Bayes' theorem.* We give only the final formula, leaving out the premises and development:

$$P(q_r|\theta \cdot p) = \frac{P(q_r|\theta) P(p|q_r \cdot \theta)}{\sum P(q_r|\theta) P(p|q_r \cdot \theta)}. \tag{1}$$

More briefly and, as Savage says[1], "somewhat informally", many authors (de Finetti, Kendall, etc.) write Bayes' theorem taking only the numerator of the formula:

$$P(q_r|\theta \cdot p) \propto P(q_r|\theta) P(p|q_r \cdot \theta).$$

---

[1] L. J. Savage *et al.*, *The Foundations of Statistical Inference*, Methuen, London, 1962, p. 15; B. de Finetti and L. J. Savage, "Sul modo di scegliere le probabilità iniziali", in *Biblioteca del Metron*, ser. C, vol. 1, Rome, Istituto di Statistica, Facoltà di Scienze statistiche, demografiche e attuariali, 1962, p. 99; M. G. Kendall and W. R. Buckland, *A Dictionary of Statistical Terms*, Oliver and Boyd, Edinburgh, 1957, p. 23; L. Faleschini, "Schema di una teoria delle scienze", in *La Scuola in Azione*, No. 12, 1st April 1963, p. 32.

The following statement derives from this:

If q_r is an event among a mutually exclusive set, conditioned by circumstances θ and p, its posterior probability is proportional to the product of the "empirical probability" or "likelihood" of p multiplied by the prior probability of q_r on θ. For some authors the prior probability would be somewhat different from the *a priori* probability of the mathematicians (section 4333), but this nuance does not alter the substance of the argument.

As an illustration, let us suppose the following:

q_r represents an individual having a height greater than 180 cm;

θ is prior information that a person belongs to the population of adult male Scots, of which the distribution by height is known;

p represents factual information that the individual has left an overcoat 150 cm long in a lobby.

On the basis of these data, the probability that the individual in question has a height greater than 180 cm, given θ and p, is proportional to the product of the prior probability that an adult Scot has a height greater than 180 cm — that is, $P(q_r|\theta)$ — and the probability $P(p|q_r.\theta)$ that an adult Scot of height greater than 180 cm would wear an overcoat 150 cm long (this last datum of probability would clearly also require a knowledge of the sartorial tastes of Scots).

If one wished to express the same situation in terms of causes (according to the classical formulation of Bayes' theorem), it would go like this. Given that the effect p (overcoat of length 150 cm) has been observed, the posterior probability that the causes q_r (great height) and θ (Scotsman) have been operative is proportional to the product of the prior probability which set in motion the cause q_r and the probability that, having set in motion that cause, the effect p has been verified (that a tall Scot wears a long overcoat).

Note that the generalized formula of the product always appears in the expression of the theorem.

433 Prior probability in Bayes' theorem

As was said earlier, Bayes' theorem has aroused lively discussion among logicians, probabilists and statisticians. While not yet concluded, these controversies challenge the validity of the theory as an instrument of inference. In truth, its application to the logical knowledge of natural events is limited by the fact that in the formulae there inevitably appears that prior probability which is a mathematical concept. When we ask, what is the probability that a proposition is true? and what is the probability that a certain premise is sufficient to justify it? or, in more practical terms, what is the probability that a human couple will give birth to a brown-haired child? and after the birth, what is the probability that its parents are homozygotic with respect to colour? — these rhetorical questions are enough to make it clear that attempts to provide answers by the use of Bayes' theorem must await a satisfactory treatment of prior probabilities, a subject which we consider later (section 4343).

4331 Criticism of Bayes' theorem as the basis of induction

Following Keynes, Borel sought the foundations of induction in Bayes' theorem[1]. But these authors, writing before World War I, are at least partially justified by the fact that it was only the new epistemology that deprived logical induction of its main support, namely the principle of the Uniformity of Nature. For this reason we shall move on to examine the opinions of more recent authors.

Harrod well understood the particular difficulty deriving from Bayes' theorem — that is, the impossibility of knowing prior probability if it is to be used in practical experimentation. This is true even in a general sense, concerning its application to the future, although, according to Harrod, this possibility exists only in the short term[2].

4332 Attempts, from Bayes to Fisher, to overcome the objections

Fisher has repeatedly stated that up to the time of Laplace (1820) the use made by Bayes (1763) of prior probability has been a source of perplexity[3] — not because Laplace rejected the concept but because he held it to be axiomatically valid. One can see here the different mental

---

[1] E. Borel, *Valeur pratique et philosophie des probabilités*, p. 37.

[2] R. F. Harrod, *op. cit.*, pp. 14–18.

[3] *The Design of Experiments*, Oliver and Boyd, Edinburgh, 1st edn, 1935. Fisher, while recognizing the importance of Bayes, criticized the postulate of inverse probability because it leads to "considering mathematical probability not as an objective quantity measured by observed frequency, but as a measure of purely psychological tendencies, and hence the respective theorems are useless for scientific purposes". The subject was taken up again and developed in recent years by the Author. See also R. A. Fisher, *Mathematical Probability in the Natural Sciences* and "The nature of probability" in the pamphlet *Smoking, the Cancer Controversy*, Oliver and Boyd, Edinburgh, 1959; also "The place of the design of experiments in the logic of scientific inference", in *Colloques Internationaux du Centre National de la Recherche Scientifique*, No. 110, Paris, 1961.

outlook of the two: the Englishman who derives his way of thinking from the tradition of Locke and Hume, and the Frenchman who aligned himself with the development of Cartesian rationalism.

Actually, on re-reading Bayes and the introductory note by Price (the editor of Bayes' posthumous work), Fisher observed that according to Bayes the *postulate* "might not perhaps appear reasonable to everyone", and that Bayes had therefore introduced an auxiliary experiment to measure prior probability, considering this artifice preferable to "introducing into his mathematical reasoning anything which could be a subject of dispute". But this avails little to widen the scope of a theorem which continues to require, in every case, an impossibility, namely the calculation of a prior probability, though it is interesting to note that the doubts expressed by later students should, at the outset, have crossed the mind of the man who proposed and proved the theorem.

According to Fisher, only rarely and in well-defined circumstances can the concept of prior probability be justified. He considers that axiomatic assumption must be absolutely excluded from it, if only because of conceptual difficulties, and affirms that no prior probability can exist when it is a question of determining certain natural measures (for example, an atomic weight). On the other hand, the concept can appear in the calculation of posterior probability, when an auxiliary experiment allows a procedure analogous to what Fisher calls the "fiducial argument" (for example, the well-known procedure of "Student", by which one sets probabilistic limits to a parent mean in normal samples, without knowing the parent variance). Fisher's attitude is unexpectedly affirmative in considering that at times (as in genetics) prior probability may be inherent in the facts themselves, although it is clear that the measure of it derives from preceding experience, and hence is none other than an inductive generalization.

4333 Justification by Jeffreys

Jeffreys[1] also retains the abundant self-confidence of the expert in a science which knows of no failures (section 38). "The principle of inverse probability", he writes, "therefore accounts for the use of the

---

[1] Jeffreys, *Scientific Inference*, p. 31.

most striking type of experiment, the *crucial test*[1] . . ., and a general theory of induction is impossible without it. Nevertheless it seems to fill many people with terror. Most statistical writers think it necessary to state that they do not propose to use inverse probability, and make an attack on the notion of prior probability. Some declare that prior probabilities cannot be assessed consistently, others that they are arbitrary, which means, if it means anything, that they can be consistently assessed in more ways than one. Some (whose logical ability in other matters is not negligible) make both criticisms. . . . Prior probabilities can be assigned consistently in many ways; it is enough, and more than enough, that all considered propositions shall have positive probabilities on *H*. One of our problems is to find the most suitable selection for our purpose." The theory of inverse probability is for the British astronomer what the theorem of Pythagoras is for geometry: without it, induction is impossible.

In order that this quotation may not unduly surprise the reader, we must credit Jeffreys with rejecting the common notion of prior probability in the mathematical sense. "The prior probability", says Jeffreys, "is often called the *a priori* probability. This is an infortunate usage because *a priori* is technically used in logic for propositions accepted independently of experience, and in probability theory would suggest that all prior probabilities were based only on the general datum *H*, used above, whereas the prior probability is intended to express simply the probability at the start of an investigation, and may have been influenced by many previous investigations. To make confusion worse, probabilities of throws of dice, which are likelihoods, are also often called *a priori* probabilities."[2]

Basically, then, Jeffreys follows the same line as Bayes, insisting on the need of ensuring that the probability generally called *a priori* does not become a purely mathematical concept, and stressing that,

---

[1] Jeffreys gives the name "crucial test of inference" to the increasing value which probability acquires by the use of formula (1) in section 4321; this value tends to unity when θ has a small probability in all the hypotheses contained in the sum in the denominator except the *r*th, and provided that the variables *q* are "comparable", i.e., that the various prior probabilities do not differ too much among themselves. Evidently, this last condition introduces the Principle of Indifference (section 44 *et seq.*), to which some theorists are opposed.

[2] Jeffreys, *op. cit.*, p. 30, footnote. The distinction between likelihoods and probabilities is clearly made, as already mentioned, in M. Boldrini, *Statistica, teoria e metodi*, p. 490.

F

on the other hand, it is a quantity based, however vaguely, on experimental information.

But — abandoning the purely philosophical aspect — one should ask, at this point, what is the presumed information which the scientist possesses when he undertakes experimental research? Ideas of regularity? Then they must be based on a principle. Simple factual notions? Then it is necessary to find a way of making concrete measures of probability correspond to them. Either way, there opens up a *regressus ad infinitum* at the root of which there will always be found an axiomatic affirmation, that is to say, an *a priori* probability.

The consequent unease is passed over by Jeffreys, despite the requirements of his work in the physical field, which are real enough when he works as a geophysicist or an astronomer. He reasons on the theory of probability without ever detaching himself from the purely mathematical terrain. This point will be taken up again later. Here we will simply say that, in another passage in the section already quoted, Jeffreys considers the case when the q_r are relatively uniform, namely a condition which violates reality and has, on the contrary, a theoretical character (cf. note 1 in section 4333).

434   Mortara's view of probability as frequency

So far we have examined the first two among the five conceptions of probability, in part derived from Carnap, whose list is given in section 42. It is now time to go on to the third, which includes those authors for whom probability can be nothing more than an experimental result. It was precisely of this that Mortara was thinking in 1922 when he defined as the "relative frequency" of a phenomenon the ratio between the number of times in which a modality of it manifested itself and the total number of observations carried out[1].

Experience shows, says Mortara, that with an increasing number of observations the relative frequencies fluctuate less and less. Thus the relative frequencies of the various numbers of the Italian State lottery, during the observed period, fluctuated a great deal more in series of 720 draws made at Palermo than did in 209,000 draws made over the whole of Italy. From this, Mortara concluded that with the indefinite increase of draws, the relative frequencies of the numbers drawn (as in every other analogous experimental phenomenon) would tend to *become uniform*. The tendency to uniformity of the relative frequencies

---

[1] G. Mortara, *Lezioni di statistica metodologica*, Leonardo da Vinci, Città di Castello, 1922, p. 415.

"is customarily expressed", adds Mortara, "by saying that the various results are *equally possible*". This factual statement the author expands into an "abstract conception" of the equal possibility of cases, from which could be deduced the type of regularity which a repeatable phenomenon would present. By characterizing the different possible cases (e.g. the balls in the lottery by numbers, or by colouring the first 45 white and the others black), one comes to define probability as "the total number of equally possible results divided into the number which are distinguished by a fixed characteristic".

It is in proceeding to the limit that Mortara sides with the mathematical outlook on probability and is thus enabled to utilize its concepts and procedures. But as regards experimental procedures, he adheres formally to the idea of frequency, of which he accepts the inevitable variations, considering them to be determined by the limited number of cases at disposal.

Apart from the interpretation which was possible in 1922, either of the results of experiments or of proceeding to the limit, Mortara's idea still appears to be substantially new. In fact, the definition of probability based on the frequencies of classified events (whether natural events or the drawing of balls) not only supplies the means for suitable historical description but also provides new instruments of forecasting, in the framework of the concept of inductive inference developed in the preceding chapter.

4341   The Kollektiv of von Mises

In contrast with Mortara, von Mises, also beginning from experimental considerations and hence rejecting the subjective concepts of Keynes and Jeffreys, explicitly axiomatizes his concept of probability.

Von Mises begins by denying that it is the function of academic philosophy to search for the true and proper essence of probability, and sets out to formulate an exact theory which could serve for the description of the sector of reality consisting of recurrent aleatory phenomena. Like Mortara, he defines relative frequency[1], and states (here recalling Poisson) that as the number of cases increases, it tends to settle down towards a limit; and he defines probability as the "limiting value of the frequency". It is a question, therefore, of an axiomatic assumption, and von Mises gives the name "collective" (Kollektiv) to

---

[1] R. von Mises, *Probability, Statistics and Truth*, prepared by Hilda Geiringer, George Allen and Unwin, London, 1955; *Manuale di critica scientifica e filosofica*, Longanesi, Milan, 1950, pp. 242 *et seq.*

the abstract series of infinite trials from which probability would emerge. The validity of the notion of "collective" appears confirmed by the fact that the rule concerning the sum of probabilities can be derived from it without contradiction. But if to the collective another axiomatic property is attributed, that is to say, the property that the members of the series appear "without regularity" (and this, basically, is the common idea of random occurrence), the product-rule of probability also receives logical justification.

4342   Castelnuovo on probability and randomness

Castelnuovo's position is unusually eclectic and thereby distinct[1]. This Italian mathematician develops the concepts of probability and frequency along parallel lines, using the former as a principle of the mathematical theory and the latter in the applications and in comparisons between theory and experience.

For Castelnuovo also, "the probability of an event is the ratio of the number of cases favourable to the event to the number of cases possible, provided that all the cases considered are equally possible (or probable)", but he hastens to add the definition of relative frequency and the consideration that, experimentally, for a very extensive class of phenomena, with an increasing number of tests the frequency fluctuates but converges towards a limit which coincides with the probability of the event evaluated *a priori*. The author explicitly warns that there is nothing absolute or rational in this, but that it is a question of pure fact which he enunciates as an "empirical law of chance".

"In a series of trials repeated a great number of times under the same conditions, each one of the possible events occurs with a frequency (relative) which is nearly equal to its probability. The closeness of the approximation increases, as a rule, with the increase in the number of tests." This is again expressed rather loosely in the following way: "The frequency of a fortuitous event in a series of experiments made under the same conditions tends, with the increase in the number of experiments, to a limit which is equal to the probability of the event."

Castelnuovo underestimates the circularity of relating limiting frequency to unknown probability; he concedes that it is possible to construct the calculus of probabilities without appeal to relative frequency, provided one does not intend to apply it in reality. He adds

---

[1] G. Castelnuovo, *Calcolo delle probabilità*, Zanichelli, Bologna, 1933, pp. 1 *et seq.*

that the laws of relative frequency sometimes help in the experimental choice between hypotheses, but finally he advises that mathematical calculus should be used essentially for forecasting frequencies, starting from probabilities, and not vice versa, i.e. using frequencies for constructing the theory of the calculus.

These are undoubtedly important observations, but insofar as they concern relative frequencies, they are valid only for describing the past, and not for forecasting the future until the laws of limiting frequencies are themselves formulated axiomatically as a working hypothesis, or until it is explicitly acknowledged that observed regularity is only one among the possible manifestations of classified regularity.

As against von Mises, who explicitly stated the need for axiomatization, Castelnuovo frees himself from this only by keeping separate the definitions of prior probability and of the law of relative frequency.

4343 Subjective concept of probability

It is not easy to interpret the thought of those writers who maintain that probability should consist of the evaluation which a man makes of the circumstances on which an event depends; but it is at once clear that they go much further than Keynes (section 43) when they claim to have eliminated, by an original approach, the embarrassing questions arising from Bayes' theorem (section 433). In other words, prior probability turns up again, but in the guise of an instrument of induction, that is, synthetic, rather than analytic as it was with the classical approach.

De Finetti, who (as mentioned in section 42) is the first and greatest exponent of the point of view we are about to discuss[1], defines probability as the subjective degree of conviction reached by a given person about the future likelihood of an event.

---

[1] Among the numerous works appearing before World War II are B. de Finetti, *Probabilismo, saggio critico sulla teoria delle probabilità e sul valore della scienza*, Naples, 1931; "Sul significato soggettivo delle probabilità", in *Fundamenta Mathematicae*, Warsaw, vol. XVII, 1931; "La prévision, ses lois logiques, ses sources subjectives", in *Annales de l'Institut Poincaré*, No. 7, 1937; *Calcolo delle probabilità*, lectures given at the University of Padua in 1937–8; "Resoconto critico del colloquio di Ginevra sulla teoria delle probabilità", in *Giornale dell'Istituto italiano degli Attuari*, 1938. With later research the subjective theory was reaffirmed by de Finetti after World War II; see "Sulla impostazione assiomatica del calcolo delle probabilità", in *Annali Triestini*, series 2, vol. 19, 1949, and *La probabilità e la statistica nei rapporti con l'induzione, secondo i diversi punti di vista*, Cremonese, Rome 1960.

The ratio between favourable chances and possible chances (probability in the sense of the classical probabilists — section 421) and that between the number of the manifestations of an event and the number of trials made (relative frequency — section 434) are, from this point of view, no more than practical expedients, useful in certain circumstances, to give a quantitative but only apparently rigorous form to the prior judgement about probability which a person is led to express about an experience, on the basis of the first ideas which he forms concerning it.

In more recent years, de Finetti and L. J. Savage, either separately or in collaboration[1], have taken the argument up again and developed it much further.

Savage states that in defining a subjective probability, both he and de Finetti are concerned with the opinions of a person as evidenced by his potential or actual behaviour. Such behaviour is determined for each individual by the fact that if he were to behave consistently, he would be inclined to bet a predetermined sum on his conviction.

Between the amount of this potential bet and the price which the same man would be disposed to pay for a purchase there exists an undoubted parallel which, according to Savage, gives a concrete meaning to the estimated probabilities. But, in fact, the bet remains only contingent, whereas the price to be paid for a wished-for purchase is something very real, and whoever makes a disbursement does so with complete awareness.

Savage insists that, after the outcome of the event (winning or losing, acquiring the purchase), the original bettable sum and the estimated price preserve a meaning, even if they have subsequently to be amended. Is it possible to establish whether the price actually paid for some purchase is more suitable than what one thought should be paid? It is said that facts speak for themselves, but they do nothing of the kind: what speaks is only the conviction one has about them, and this is equally true of an estimated probability as of a determined price. That this is really the case is shown by the satisfaction or annoyance that one feels after a purchase, when realizing either that one has spent less than expected, thus enjoying an unexpected gain, or that one has

---

[1] B. de Finetti and L. J. Savage, "Sul modo di scegliere le probabilità iniziali", *loc. cit.* L. J. Savage *et al.*, *The Foundations of Statistical Inference* (quoted above). See also H. Kyburg, *Logical and Fiducial Probability*, presented to the 34th Session of the International Statistical Institute, Ottawa, 1963; M. Boldrini, Presidential Address, presented to the 34th Session of the International Statistical Institute, Ottawa, 1963.

been cheated and has lost money: in short, it is the old idea of "consumer's return" applied to the concept of probability.

4344 Mathematical formulation of subjective probability

The transition from these general ideas to the mathematical foundation is effected by the introduction of what de Finetti calls the "principle of consistency". It will be remembered that, in the jargon of the gaming table, the "mathematical expectation" of a chance win is the sum which one is willing to bet on the hoped-for win. Call p the mathematical expectation when the sum to be won is unity. p will also be the degree of faith placed in the favourable occurrence of the random event. The principle of consistency states that it should not be possible to make a bet such as to assure to the player a certain win. Consequently, the mathematical expectation p must not be less than zero, because otherwise the player would be paid to bet, and it must not be greater than 1 because he would be betting, at any coup, more than he could win when successful.

With these elements and a few other principles, it is possible to transfer the concept of mathematical expectation relative to bets such as the above to that of probability p, and from this to develop the theorems proper to this branch of mathematics.

The point is that the player has a free choice to evaluate the probability p in the same way as he would evaluate a mathematical expectation, in sequences of random events. Such a free choice, as has been said, is limited by the fact that p is necessarily contained between 0 and 1 (inclusive). In evaluating p, the probabilist, like the games player, will do his best to gather the most correct information possible; and then both, having made their choice, work subjectively.

4345 Concentrated and continuous probability

The foregoing considerations appear to conform to good sense: and in part they do. But the same mathematical formulation also reveals their practical limitations.

There are certain mathematical expectations or probabilities to which one can attribute, with great plausibility, a value of either zero or one: and it is precisely in these cases — called by Savage and de Finetti "concentrated" probability — that subjective probabilistic evaluation can be effected without risk. Savage exemplifies these with the criterion of measure now known as "specific gravity", which was first suggested by Archimedes for checking the suspected adulteration of the gold for Hiero's crown; and he justifiably points out that to the king's conviction of not having been swindled there corresponds the probability zero (or almost) that the gold might contain a foreign metal in appreciable proportions (not in negligible proportions, because the means were lacking then, as now, of excluding, for example, the presence of at least one atom of another metal). Similarly, the probability that heat or cold beyond certain extremes can kill is equal to 1, or very nearly: and one keeps an open mind on certain remote possibilities, such as the survival of Daniel's companions in the furnace of Nebuchadnezzar, or that of salamanders supposedly frozen for thousands of years.

In the contrary case, when probability is "continuous" or spread over a range, there is considerable latitude of choice in assigning prior probabilities. Consequently, in these circumstances, an attribution of probability without adequate measures would be entirely unacceptable: the subjective method would be inapplicable here.

On a higher level of thought, one can formulate a similar problem about the cathartic function attributed by St Paul to Christianity (Ephesians, 4, 20–24). To the old self "which is corrupt according to the deceitful lusts", it is not possible to attribute any degree of sinfulness in the absence of observation, because this can vary from zero to one, while it is permissible to consider as pure, and hence as a sinner with probability zero, the new self "which, after God, is created in righteousness and true holiness".

These and similar examples lead one to think intuitively that in practical research the case of concentrated probability is exceptional, while that of continuous probability is the rule. It follows that probabilities relating to measures cannot usually be evaluated *a priori*, even in a largely presumptive way, without having arbitrary recourse to the so-called "Principle of Indifference", which is always justifiable axiomatically (section 423) but in experimental research is useful only under specific conditions (section 44 *et seq.*)[1].

4346   Subjectivity in science and probability

The difficulty of evaluating probability subjectively when its scale varies between zero and one (end-points excluded) arises from the differing opinions to which people are prone. De Finetti is correct in not regarding this as an objection of principle, but he should likewise

---

[1] Kyburg, mentioned earlier, writes in similar terms: "No general principle for measuring *a priori* probabilities has up till now been accorded anything like universal consent, nor does such a principle seem suited to obtain it".

recognize that, to overcome this obstacle, one has to rely on methods of reconciling different opinions, e.g. by considering the distribution of individual differences, or by recourse to experimental results.

Savage, instead of solving the difficulty, tries to get round it by generalizing. He points out that the usual applications of the probability calculus to scientific investigation, though called "objective", are, for the most part, not so. He points out, for example, that anyone familiar with experimental variation is willing to postulate that a variable x is normally distributed around its mean μ with a variance 1, or that x is a Poisson variable with a mean (and hence a variance) σ, or that x is distributed normally with a mean μ and variance σ.

These probabilities are not in fact objective but expressions of opinion. In practice, they are hardly ever seriously accepted as realistic hypotheses by experts, but are considered rather as a rough-and-ready way of approaching given problems, until such time as more valid assumptions become necessary. For example, if x represents 15 physical measurements with presumed normal distribution, serious doubts will arise if the largest value lies well away from the others. This means that the investigator is uncertain about the assumed normality of the distribution, even if he does not say so. In spite of this, not even Savage goes so far as to exclude the need of choosing initial assumptions, on the understanding, however, that they are provisional and remain valid only "until more realistic assumptions become necessary and possible".

De Finetti confirms that special positions are often accepted "in which certain kinds of information (asymmetrics, frequencies, etc.) appear so weighty as to lead to a crude identification of probability with frequency, or with a ratio between favourable and possible cases, without one's realizing that this decision, too, constitutes nothing more than a form of personal evaluation".

An attempt at rebuttal of these remarks was mentioned above. Indeed, they are inherent in the general principles of epistemology; and if the revealing of subjective elements were in fact sufficient to threaten the basis of science, not only the concept of probability but all natural knowledge would be weakened. But what, then, constitutes "objectivity" and "subjectivity" in science? The answer is readily found by considering some of the variable random elements in physical measurements.

Whether a body is hot or cold is a matter of opinion. Apart from the extremes (already considered) which kill, the heat or cold of an external object is merely a question of personal sensibility; and this is so difficult to communicate that — for example — on a spring day one

F*

person will go out wearing an overcoat and another a jacket or perhaps in shirtsleeves. This disparity and the consequent uncertainty are avoided by the universal convention of relying on the thermometer to measure the temperature. The initial arbitrary choice is between the scales of Celsius, Fahrenheit or Réaumur. After that, temperature readings become objective chance variables, communicable and comparable, which do not in the least make people regret their personal judgements about the temperature. This does not mean losing one's awareness that the "objective" measures of the thermometer do not correspond to "subjective" sensations and do not influence them. It only means accepting, once and for all, an arbiter to evaluate the temperature.

An analogous situation occurs in regard to the determining of duration (section 233). In science one speaks of sidereal, biological and psychological time. We know that for centuries durations were — and for many people still are — evaluated on the basis of the variable interval between the rising and setting of the sun[1]. But at a certain stage of social development the urgent need arose to adopt uniform — that is, objective — measures. And as soon as the laws of the pendulum were noted, men started to make very good clocks; and by these today they can measure the hours, sad or gay, boring or busy. And what is to be said about distances, which used to be judged subjectively (and sometimes still are) as "a rifle-shot away", "a morning's walk", "a day's camel-ride", but have now become standardized and are generally understood whether transformed into miles, metres or parasangs?

From these considerations one concludes that all human experience is subjective, and consequently science must forget its impossible aspirations about the "thing in itself"; but, at the cost of self-abnegation, it cannot ignore objective phenomena or, when possible, their quantitative determination.

Thus, when the subjective idea of probability is submerged in the great ocean of human evaluation of the facts of Nature, and when there is a clear need of making evidence uniform and intercommunicable by using a scale, one does not (for example) have to go out of doors, if there is a thermometer outside the window, before deciding whether to put on an overcoat. The probability that one is hot or cold is written on the thermometer and one acts accordingly.

---

[1] ". . . I said to the master that he should stop and wait for me for the eighth of an hour", wrote Benvenuto Cellini in the *Autobiography*, XLVI, about the year 1557, when there were as yet no reliable clocks. This is an unusual and completely subjective measure of time.

4347 Subjective probability of historical entities; objective probability of classified events

It is appropriate here to draw attention to a different formulation of the concept of probability, which in one respect refers to its experimental basis and in another brings to light a possible subjective aspect. It has already been shown (section 354) that the world M is the union of all single historical entities H, or (what comes to the same thing) of all events classified as S. Using the symbol $\cup$ to indicate aggregation or "logical sum", Professor Faleschini defines the world as follows[1]:

$$\cup H = \cup S = M.$$

Given this premiss, to every aggregate of type S (accidents, herds of Friesian cows, economic crises) there corresponds a measure in probability P; and since individual events are mutually exclusive, the probability that one of them will appear necessarily tends to unity in virtue of the additive property of probabilities. Supposing, therefore, that $P(S) = 1$, we shall also have $0 \leqslant P(E) \leqslant 1$, where E is any subset of S, and hence $\cup P(E) = P(S) = 1$.

On the other hand, the probability of a definite historical object H (Tower Bridge, Vostok III, etc.) which, on the human scale, is practically unrepeatable, will be $P(H) \simeq 0$. A historical fact is actually constituted by the convergence of a group of characteristics mutually linked together, each one of which belongs to a scientific aggregate. Its probability is therefore of the order of magnitude of the product of numerous probabilities (one for each characteristic), and hence is very near to zero. Although the probability of some bridge or other is high, the probability of this particular bridge, *in all its detail*, is small.

Given a historical situation H, suppose that it can evolve into situations H' or H'', which are mutually exclusive. In such problems there is point in evaluating $P(H'|H)$ and $P(H''|H)$, that is, in attributing a value to the probability that the situation H may evolve towards H' or towards H''. By a general theorem of probability, using the sign $\cap$ to indicate the intersection, we have: $P(H'|H) = P(H' \cap H)/P(H)$. However near to zero $P(H)$ may be — and also, *a fortiori*, $P(H'|H)$ — it may be possible to attribute a meaning to some value of their indeterminate relation $0/0$, especially if such a value is near to zero or to one. An evaluation of this kind can give useful practical results because

---

[1] L. Faleschini, *Schema*. For the definition of the world, see the first axioms of Wittgenstein (section 141): "the world is the totality of facts, not of things" (axiom 1.1); "what is the case — a fact — is the existence of states of affairs" (axiom 2).

it allows one to forecast future reality, and hence to exercise a choice. The probability selected from among the values of the indeterminate relation 0/0 could only be of a subjective nature, because it depends on evaluating the information that the person making the choice can collect about the structure of the aggregates H and H'.

On the other hand, the value which can be attributed to $P(E)$ always remains defined on the basis of experimental results and hence deserves to be described as an objective probability. The usefulness of these distinctions will emerge later (section 5733).

4348   Pragmatic justification of the objective theory

Turning back to the points dealt with in section 4345 and to those developed above, we do not mean that discussion about the subjective or objective nature of probability is useless; on the contrary, we give point to it by recognizing its consistency with the general principles of epistemology. De Finetti, Jeffreys, Savage and other mathematicians who, at least in part, profess similar ideas and theory, are concerned above all with defending the utility of Bayes' theorem. This contains an experimental probability (always measurable, at least in theory) and an embarrassing and unknown prior probability (sections 421 and 4321). If this last item could be estimated in advance, the product of the two probabilities would immediately give the so-called probability of the cause. However, most people affirm that subjective evaluation is impossible or too hazardous, and in any case differs from person to person; for this reason they reject Bayes' theorem as a practical instrument. From the opposing position, the subjectivists proclaim it as a very fruitful theorem, while some, such as Jeffreys, consider it an indispensable instrument of induction.

We have seen earlier that Harrod and many others reject this latter point of view. For me, the question is a little more complex. Undoubtedly, the theory of induction developed here, by which any general premiss assumed from outside is rejected, does not in fact require reference to Bayes' theorem. On the other hand, the above considerations on the subjective root of all measures corresponds to the general thesis that science is exclusively a human product in which the subjective element constitutes the weft, which is woven on the warp of the external world; and hence this weft is also operative in the particular field of probability.

But it is one thing to recognize the vast area of subjectivity in investigations of Nature, and another to stop there and to limit the use of these linguistic instruments (measures) which, as we say, render science objective and submit it to predetermined rules which ensure the intercommunicability of the results among all participants. Hence the conclusion — agreeing with those of most authors we have quoted and of others we have not, and recently re-affirmed by Bartlett in discussion with Savage[1] — of not preferring "types of so-called subjective probability or 'degrees of belief' ."

435   Carnap's view of probability as degree of confirmation and as relative frequency

To come to the last heading of the classification of probability theory given in section 42, it is evident that in his axiomatic system Carnap takes care to distinguish between undemonstrated propositions, which he calls "explicanda", and demonstrated ones, which he calls "explicata". Naturally an "explicandum" is a weak basis for any reasoning, but its validity is strengthened once it is transformed into an "explicatum". Propositions relative to "probability$_1$", to be understood as "degree of confirmation", and to "probability$_2$", to be understood as "relative frequency" (compare section 42), are both simply "explicanda". Carnap regrets that, because of the general indiscriminate use of the word "probability", insufficient attention has been paid until now to these two meanings, which are distinguishable even though they are inductive in nature and although they overlap. This recalls the discussion in section 1421 on the confusion arising from the use of the same word for the two ideas of "electron as wave" and "electron as corpuscle", and also for the ideas of "water" (section 3553) and of "time" (section 366). Carnap adds that suitable synonyms exist in many languages, such as the Latin terms *probabilis* and *verisimilis*, which have their corresponding words in French and Italian, and the English terms "probable" and "likely". But these have been used so indiscriminately that, to obviate further confusion, Carnap thought it necessary to avoid using them at all with special meanings.

By probability$_1$ (that is, "degree of confirmation") is to be understood the logical judgement expressed by an observer after gathering all the relevant elements about a factual hypothesis H and working these into a statement P, so as to decide whether and to what extent the hypothesis itself is reliable. Carnap explains that confirmation in this sense, being founded on concrete data and expressed in linguistic form, is a "semantic" datum (cf. section 144) — as is often the propositional notion of probability given by other writers already con-

---

[1] L. J. Savage *et al.*, *op. cit.*, p. 37.

sidered. It has an inductive character even when one refers to facts only thought of but not yet experienced.

"Probability$_2$", on the other hand, expresses the relation between two classes of events. Suppose that in 6,000 throws of a die the one comes up 1,020 times. The ratio $1 \cdot 02/6$ can be considered as the probability of throwing a one with this die. It does not coincide with the theoretical number $\frac{1}{6}$, adopted by the mathematicians, which is purely abstract (section 423); the differences can depend on the chance outcome (for the moment it cannot be called otherwise) of the games with which mathematical statisticians are concerned, and also on the imperfections of the particular implement[1].

The result of the experiment should be interpreted as follows. Between the class of events "throws with this die" (indicated by K) and the class of events "turning-up of a one on the die" (indicated by N) there exists the numerical ratio furnished by the series of trials carried out and considered sufficiently extensive. Naturally, the adverb "sufficiently" is also susceptible of mathematical evaluation.

In sections 42 et seq. we noticed that some mathematicians and statisticians (von Mises, Reichenbach, Wald, and perhaps Mortara) conceive probability$_2$ as the limiting frequency in an infinite experience. As against these, other writers, such as Castelnuovo and R. A. Fisher, are nearer to the idea of that probability$_2$ in which Carnap finds "an undefined term in an axiomatic system". The class of reference K is called a population or universe, and is not necessarily enumerable, as it can concern a continuum.

44   The principles of Indifference and Indeterminacy

From the thoroughgoing investigation of the logicians has emerged a critical analysis of the Principle of Indifference, which is the basis of the classical formulation of probability theory. A further criticism of the fundamentals of the classical doctrine turns on the concept of Universal Causation which, under the guise of probability, is today viewed obversely, that is, as the so-called "Principle of Indeterminacy". Both these questions must now be faced.

---

[1] The distinction recalls a further measure of probability, characteristic of the implements with which probability is tested, and which is called $_0p$ in M. Boldrini, *Statistica, teoria e metodi*, 1962 (1st edn, 1942), pp. 273, 490 and elsewhere. This could well be put alongside Carnap's p_1 and p_2.

441 Keynes's conditional acceptance of the Principle of Indifference

Keynes points out that for a proposition of whose truth or falsity we have no knowledge, it would be absurd to state that the probability of the two alternatives is $\frac{1}{2}$.[1]

Keynes recalls that Boole had already objected that he could not attribute the probability $\frac{1}{2}$ to the case of perfect ignorance — but rather, if anything, the indeterminate expression $0/0$, though he immediately adds that this remark applies at most to the predicate, never to the subject.

For Keynes, the Principle of Indifference "certainly remains as a *negative* criterion; two propositions cannot be equally probable, so long as there *is* any ground for discriminating between them. The principle is a necessary but not, as it seems, a sufficient condition." It can be accepted only in the case where no valid reason suggests a preferable principle.

One speaks of "indifference" or "preference" between two propositions q and r on the hypothesis p, according to whether $P(q|p) = P(r|p)$ or $P(q|p) \neq P(r|p)$.

Furthermore one speaks of the "irrelevance" or "relevance" of the circumstance p_1 for q according to whether $P(q|p) = P(q|p_1 \cdot p)$ or $P(q|p) \neq P(q|p_1 \cdot p)$.

Clearly it is unnecessary for a circumstance p to be relevant for q in order that one can speak of indifference between q and r on the hypothesis p. Keynes rightly states that it is only this proposition, subordinating indifference to irrelevance, that permits us to justify the Principle of Indifference without getting into a vicious circle. Undoubtedly this somewhat limits the field of application of the principle, though without quite eliminating it — rather does it strengthen its applicability.

Keynes shows at some length how, by complying with the conditions of relevance, the antinomies of many practical cases described by logicians and mathematicians can be cleared up. "Most of the examples", he says, "to which the mathematical theory of chances has been applied, and which depend upon the Principle of Indifference, can

---

[1] Cf. J. M. Keynes, *A Treatise on Probability*, see especially Chapter IV, pp. 41–64; G. Boole ,*op. cit.*, pp. 89 *et seq.*, 263 *et seq.*

be arranged, I think, in the forms which the rule requires as formulated above."[1]

4411　An application of Keynes's concept

A subtle example, to which Boole and many other mathematicians refer, is that of the urn containing black and white balls in unknown number and proportion. Two hypothetical solutions present themselves. The first supposes that half the balls are black and half white, and hence that the probability of drawing a white ball is $\frac{1}{2}$. This solution does not take into account the "relevant" fact that, in reality, the ratio can be very varied (influenced as it also is by the number of balls) and that it is modified by each extraction not replaced.

There is another more convincing hypothesis by which, says Keynes, the Principle of Indifference finds a justified application. That is, suppose that each single ball may be either white, w, or black, b. In the particular case of only two balls A and B, the hypothesis allows that we may have A_w or A_b and also B_w or B_b; and similarly if the number of balls is more than two.

Clearly, such a "constitution" of the urn admits of only one hypothesis under which there is numerical equality between black and white balls and, what is more, it considers the probability of a white ball always "logically" constant and equal to $\frac{1}{2}$, even after one or more balls have been drawn from the urn and not replaced.

On the other hand, in the problem in which a card is extracted from each of two packs, proposed in section 422, three instances of the application of the Principle of Indifference present themselves, all equally unjustified. In Poisson's solution the principle is invoked with regard to the three ways of pairing the suits (black–black, black–red and red–red); in von Kries' solution, it is applied to the single cards of the two packs; finally, in the uncertainty of choice in which Poincaré finds himself, the principle produces two solutions, neither of which can be said to be right or wrong.

442　Harrod on Carnap's treatment of the Principle of Indifference

The references in sections 42 and 435 to Carnap's system of probabilities are enough to make it clear that in his axiomatics there is no room for the Principle of Indifference; one would conclude that he

---

[1] Kyburg, already quoted, agrees with Keynes that the Principle of Indifference is not self-contradictory, but insists on considering it as completely arbitrary.

had not concerned himself with it in his *Foundations*. Harrod recognizes the intention here, but suggests[1] that the principle had crept back, unperceived, in a refined form; Carnap's effort would seem to be directed towards choosing alternatives with equal prior probability. According to Harrod, this consideration — actually somewhat obscure — leads one to think that Carnap was seeking a logical justification of the choice, in the manner of the old classical probabilists. The justification put forward by Carnap consists partly of a somewhat vague reference to likelihood and simplicity, and partly of his confidence in being able to derive series of inductive principles conforming to those in general use.

As to Jeffreys, it was pointed out in section 4333 (note) that the crucial experiment of inference requires that only one of the causes θ, which figure in Bayes' formula, should be large and that the values of q_r should be relatively uniform. This last condition certainly constitutes a frank appeal to the Principle of Indifference, but there is nothing surreptitious about it, since the author openly upholds the classical rules of induction.

4421 Carnap on the Principle of Indifference

We now consider Carnap's views about the logical significance of the Principle of Indifference; his attitude is other than what might be expected. He begins by affirming that, of the two definitions of probability formulated by him (both of an inductive character), mathematicians have for the most part accepted only that which he specifies with the suffix 1, that is, the "degree of confirmation of a proposition", and not in fact probability$_2$, which expresses "relative frequency". In the cases nearest to those considered by mathematicians, probability$_1$ "is determined with the help of a given frequency, and its value is either equal to or close to that of the frequency". When the frequency is established as if it reflected the entire experience, it is called prior probability. This expression, explains Carnap, "was used in cases where the evidence did not state a frequency but was very weak or even tautological (a 'statement of ignorance'), and the value of the probability was determined chiefly by the use of the Principle of Indifference".[2]

Thus in Carnap's system we seem to have a distinctly negative position. But is this really so? Consideration of the weak character of

---

[1] R. F. Harrod, *op. cit.*, pp. 21–3.
[2] R. Carnap, *Logical Foundations of Probability*, pp. 188, 343.

the "explicanda", will give a different answer (section 435).

"At first glance, the classical theory seems much stronger than modern axiomatic systems. And we find, indeed, many stronger theorems stated by classical authors. However, closer examination shows that the proofs of these stronger theorems make use, explicitly or implicitly, of the Principle of Indifference. Classical theory claims to give a definition of probability based on the concept of equipossible cases. The only rule given for the application of the latter concept is the Principle of Indifference, and since we know today that this principle leads to a contradiction, there is in fact no definition of the concept of equipossibility."

Carnap, like Keynes, changes his point of view and emerges on a much freer level. "In order to base the classical theory on a consistent foundation we may proceed in the following way. We regard it, not as an interpreted theory, as was intended, but as a system of uninterpreted axioms with 'equipossible cases' assumed as an undefined primitive term without interpretation; that is, we accept the classical definition of 'probability' based on the 'equipossible'. Thus, the definition becomes an uninterpreted axiomatic definition. If we do this (and, in addition, make some other necessary modifications, e.g. by inserting references to evidence, which are often omitted in classical formulations), we obtain a consistent system of axioms. This system, however, is as weak as the modern systems already described." Thus, perhaps to ensure consistency with his own axiomatic system, Carnap ends up by justifying the assumption of indifference made by the classical probabilists.

443   Harrod's rejection of the Principle of Indifference

Harrod, unlike the authors mentioned so far, is strongly opposed to the Principle of Indifference. The classical probabilists, he says, have found a convenient pretext to free themselves from the state of total ignorance in which man has been placed by Nature. Their "proposal has had various names — 'the Principle of Indifference' or the 'Equal Distribution of Ignorance'. It is totally unacceptable."[1]

It is not possible here to summarize all Harrod's arguments, some of which are obscure, and we must confine ourselves to a few significant points.

Suppose we have an extremely exact die: will the probability of

---

[1] R. F. Harrod, *op. cit.*, p. 18.

throwing a one be equal to $\frac{1}{6}$, as mathematical theory postulates? Following Carnap's reasoning, we reply that it depends; that probability$_1$ is one thing, and probability$_2$, which includes the particular quantity $_0p$ expressing the correspondence of a material gaming implement with theoretical ideas, is another (cf. section 435, note).

Harrod maintains that, despite the most accurate technique, there remains necessarily a margin of error in the die[1]. Suppose the odds are 999 to 1 in favour of the die being unbiased within limits of tolerance so fine as to be negligible. Then the probability of throwing a one is at least equal to $\frac{999}{1000} \cdot \frac{1}{6}$. The probability of throwing one of the other numbers is at least $\frac{999}{1000} \cdot \frac{5}{6}$. The total of the two fractions $\frac{999}{6000}$ and $\frac{4995}{6000}$ is not equal to 1; consequently, in contradiction with the definition, they do not represent true probabilities. But, adds Harrod, in spite of theory, the die will never remain upright on an edge or a point[2]; hence the phrase "at least".

What, then, is the true probability? One might be tempted to split the difference between 1 and $\frac{5994}{6000}$ equiproportionally between the six facets, after which the two fractions would become $\frac{1}{6}$ and $\frac{5}{6}$, in accordance with mathematical theory. But this recourse to the Principle of Indifference would be entirely unjustified.

Clearly the difficulty is not overcome, rather is it aggravated, if the probability is measured not on the throw of one die but of a population, say 1000, of very accurate dice. It is certain that in such a case the imperfections would be spread over the different faces. But there would always remain some imperfection, however small, to be eliminated by recourse to the Principle of Indifference. Furthermore, the new principle would have to be introduced which mathematicians call the "compensation of errors".

4431 Limits to objections to the Principle of Indifference

Harrod's objections, however, seem to bear not so much on the Principle of Indifference as on the incautious application of it. Keynes makes the same point. Carnap's development maintains that prob-

---

[1] *Ibid.*, pp. 44–5.

[2] This reasoning is not too persuasive because one need not rely on the Principle of Indifference in order to ask oneself whether the probability of throwing a number other than a one is not the complement to unity of the first fraction written above, i.e. $1 - \frac{999}{1000} \cdot \frac{1}{6} = \frac{5001}{6000}$.

ability$_1$, which causes the mathematicians to affirm that it is $\frac{1}{6}$ for each face of a die, has only an axiomatic basis; it is an "explicandum" not a fact, and as such has value for the purpose of deductive developments. Harrod, on the contrary, works exclusively on the inductive level and should accordingly justify his rejection without blaming the classicists (who indeed do not come into it) for their different attitude. To confuse the two attitudes would be like confusing the geometrical properties of a sphere with those of the terrestrial globe, which, in ordinary language, is said to be round. The example helps one to sense the value of systems — and it will be seen that the Calculus of Probability is one of them — for interpreting Nature. Suppose the surface S of the earth to be known exactly and its radius r to be properly estimated. The fact that one may meet with an inequality $S \neq 4\pi r^2$ will be interpreted neither as an error of geometry nor as a rebellion of the earth against the principles of Euclid, but will simply testify to the formula being unsuitable to represent the shape of our planet. Analogously, a deviation of one six-thousandth from the supposed behaviour of a perfect die would merely show that a mathematical axiom and the experimental results can exist side by side.

444 Conditional validity of the Principle of Indifference

We need only recapitulate some of the concepts on induction set forth in Chapter 3 to recognize that the validity of the Principle of Indifference can be maintained, provided that the cautions with which Keynes surrounded it are observed and that it is not applied to artificial logical situations invented to discredit it.

Above all, let us be clear that in a deductive system, such as mathematics, to postulate that principle would be permissible even if in the practice of the factual disciplines the opportunity never occurred of invoking probability as the instrument of systematization and discovery. The Calculus of Probability, on the contrary, has moved into a more prominent position; having come into being around the gaming table, from the start it has had to face up to practical problems. And it is universally known that all the natural sciences, from statistical mechanics to combinatory philology, have continual recourse to it in their practical difficulties (see sections 445 *et seq.*).

At the end of Chapter 1 it was emphasized that, along with abstract mathematics, another quite similar discipline flourishes, which makes use of the same algorithms and follows the same deductive processes, but in which the quantities are not empty mental constructions but rather practical, natural events; the conclusions of the theorems are not tautologies but interdependences, facts and not concepts. Its nature will shortly be discussed. But here, taking the usual questions arising from games of chance, we are referring to Carnap's second idea, that of probability$_2$ — very often agreeing with probability$_1$ — which refers to the relation between events; for example, to the outcome of repeated throws in games of chance. The question presents certain difficulties which cannot be dealt with here, but the formal relation between "mathematical probability" and "empirical" or "experimental probability" or "relative frequency" is very obvious. When we say that the probability of throwing a one with this die — and only this die — is empirically equal to $\frac{999}{1000} \cdot \frac{1}{6}$, we mean that in repeating the experiment many times the number 1 appears about 166 times in 1000 throws.

If one did not expect something very similar to continue, this relation would express an historical fact, not a probability. But there is the empirical expectation (although not logically justifiable) that, in a second series of 1000 tests carried out with the same instrument and under identical conditions, the 1 will appear a different number of times, but not very different, from 166; and it will then be a question of evaluating the discordance between the two results and those expected from a further series of tests.

4441   The Principle of Indifference in theory and in games of chance

One cannot say that mathematicians were not aware that their position differed from that of the first players who supplied the impetus for their researches. As will appear later, this position tallies with that of the statisticians (cf. heading (c) in section 42). One need but go back to Diderot's (section 423) instancing events "which we know or suppose" to have the same possibility of happening; to Laplace (*ibidem*) who speaks of "reducing all events of the same kind to a number of equally possible cases"; and finally to the opinion of Carnap (section 4421) who affirms that probability$_1$ in the sense nearest to that of mathematicians "is determined with the help of a frequency, and its value is equal or near to that of such a frequency".

Fisher (section 41) has the right idea when he says that the mathematical calculus has been facilitated by the construction of excellent gaming implements. This is certainly true, because mathematicians, finding (for example) that the 1 of a die turns up with frequencies varying little and contained in the limits $(x \pm \Delta x)/6x$, have been able to round off to $\frac{1}{6}$ the probability of the 1 in general. They could thus make the two terms of the fraction coincide with the relation of the single 1 to the total number of faces of a hypothetical die, assumed to be perfect.

4442 Bernoulli's theorem

But we can go further. One of the basic theorems of the calculus of probability, called the "law of large numbers", shows that the probability of a deviation between probability and frequency tends to a limit as the number of trials increases. James Bernoulli worked on this theorem for years, and it has earned him undying posthumous fame.

The modern way of enunciating Bernoulli's theorem is as follows. Given an event F whose probability remains constant and equal to p, and a positive number chosen at will as small as one likes, one can carry out a number of trials n, large enough to make the probability of the inequality $\left|\dfrac{r}{n}-p\right| < \epsilon$ tend to certainty, r being the number of successes. In other words, ϵ becomes as small as one wishes, while n is increased at convenience.

This theorem — in spite of its author's aspiration to apply it in practice — is purely hypothetical and is concerned with abstract experiments which are assumed to be regulated by a mathematical probability p. However, it did not remain unfruitful, because the earlier probabilists had already convinced themselves experimentally of the existence of a practical relation between probability and frequency in games of chance. Some examples have been given from the game of lotto, recalling (section 434) Mortara's concept of probability. A further confirmation can usefully be added. Two series of draws from equal numbers of black and white balls gave the following results:

n	Number of white balls		Difference between p and $\|a-b\|/n$
	Calculated on the basis of $p = \frac{1}{2}$ a	Observed b	
200 draws	100	93	0·035
1,000 draws	500	493	0·007

The diminution of the relative deviation is very marked and, despite the absence of a logical link, seems to accord with Bernoulli's theorem.

4443 Application to games of chance

A deeper appreciation of the problem is obtained by considering the outcomes of numerous series of draws. For simplicity we will refer to the results of 10 draws repeated 20 times from a bag containing black and white balls in equal number. They are, in fact, the 200 draws of which we spoke. Clearly, in a series of draws the white ball can appear 0, 1, 2, . . ., 10 times, and hence the black one can appear 10, 9, 8, . . ., 0 times. Actual trials produced the following totals:

White balls	Black balls	Frequencies in 20 trials	
		Actual	Theoretical
0	10	0	0·02
1	9	1	0·20
2	8	0	0·88
3	7	3	2·34
4	6	5	4·10
5	5	6	4·92
6	4	3	4·10
7	3	1	2·34
8	2	1	0·88
9	1	0	0·20
10	0	0	0·02
		20	20·00

Ten draws in each set are clearly not many, and yet an equal number of black and white balls was obtained in six sets out of 20, which consequently appears to be the most frequent result. The adjacent numbers were a little less frequent — five sets with 4 white balls and three sets with 6 white balls. These results all seem to accord roughly with the theoretical ones, calculated on the basis of the supposed probability $p = \frac{1}{2}$, in the last column of the above table.

In fact, known theorems of the calculus of probability show that in theory (though not necessarily exactly in practice) the maximum frequency corresponds to the probability, and the frequencies diminish as we move away from the maximum. This is in accordance with the theorem of Gauss–Quetelet, referred to in section 381.

445 Classificatory homogeneity and the Principle of Indifference

Let us consider the throwing of a die and turning up, say, a 6. It is evident that the result can be obtained in various ways, such as the die falling from different heights or rolling several times, and having the spots differently orientated to the player; similarly for all the other numbers. All this is said to be a matter of indifference. We are evidently concerned with unequal entities which only conventionally become homogeneous events and can thus be counted, expressed in relative frequencies, and idealized in mathematical probabilities. This remark, perhaps over-analytical, shows that a "success", a "failure" and a "relative frequency" in throwing a die are common nouns, constituted of homogeneous events — all equally abstract, although not empty of content, and perfectly comparable with the quantities which represent mathematical probabilities (cf. note 2 to section 42). When we speak of "a case" we are not describing the particular way in which a specified face of a certain die presents itself at a given time in well-defined circumstances, but we mean the abstraction of some characteristics which make it comparable to other cases; when we speak of "relative frequency" we are using a common noun in no way unlike "oak" or "molecule of water" (cf. sections 354 *et seq.*).[1]

Therefore an experimental probability is an induction; it is based on homogeneous factual data which can result from throwing dice or playing cards or from the numbers of a tombola, but being concerned with classified facts and not with axiomatic propositions it could have any content whatsoever.

If someone entered a wood of equal proportions of birch and oak and selected a tree, he would have a probability of $\frac{1}{2}$ of picking an oak or a birch. In the same way, if he put as many black balls in a bag as there were oaks in the wood, and as many white balls as there were birches, he would have a probability of $\frac{1}{2}$ of drawing out a black ball or a white.

The way in which the notion of probability is applied to natural facts is therefore extremely simple, because one assumes circumstances which allow recourse to the Principle of Indifference. The advice of Laplace, to reduce all events to equally possible chances, seems to

---

[1] For a scholarly use by Galileo of simplification and homogenization by classification in the theory of motion, see the first part of the volume by H. Butterfield, *The Origins of Modern Science, 1700–1800*, G. Bell and Sons, London, 1949, pp. 67–82.

allude precisely to the ordinary mental process of transforming a series of historical experiences into common nouns.

4451 The Principle of Indifference applied to classification of the stars

One could hardly adduce a more characteristic example of the purely conventional meaning — necessary rather than merely useful — of the Principle of Indifference than the classification of stellar magnitudes. Since the time of Ptolemy, the stars visible to the naked eye have been grouped in six magnitudes according to their luminosity. Those which, before all the others, scintillate in the dusk of the evening and disappear last in the morning are classed as of the first magnitude; those of the sixth magnitude are the faintest, which a normal eye can scarcely distinguish on a clear moonless night, while the second, third, fourth and fifth are intermediate. In modern times the criteria of classification have been refined, and a region of the sky near the North Pole has been selected for observation. The unit of measurement for luminosity is the Pole Star, and the equipment consists of photometers and other instruments. But the old criterion, although built up in this way, persists unchanged.

Now all the stars visible to the naked eye have been distributed in the six classes, and the stars in each class have, conventionally, an apparently identical luminosity. It is clear that, given the 5719 stars visible in the whole sky, divided, according to Houzeau, into magnitudes (I = 20; II = 51; III = 200; IV = 595; V = 1213; VI = 3640) and supposing them to be evenly distributed, someone pointing casually at the sky has a probability, for example, of 595/5719 of indicating a star of the fourth magnitude. If the calculations are correct, this is unexceptionable; all the chances are equally possible. But how can this be? Only because man has placed in class IV, considering them all of the same luminosity, those 595 stars, not one more nor one less.

4452 The Principle of Indifference in quantum theory

It is very important to notice that probability is also conceived in exactly this way in the quantum theory of radioactive phenomena. According to earlier mechanical formulations, the field of force surrounding an atomic nucleus should prevent the particles from escaping from the nucleus itself. On the contrary, affirms Bohr[1]; "in quantum

---

[1] N. Bohr, *Teoria dell'atomo e conoscenza umana*, Boringhieri, Turin, 1961, p. 370. This collection of papers was expressly prepared for the Italian edition, under Bohr's supervision.

mechanics the situation is different in that the field of force constitutes an obstacle which, for the most part, holds back the material waves but permits a small portion of them to escape. That part of the waves which gets out in a certain interval of time gives a measure of the probability that, in that interval, the atomic nucleus will disintegrate." The concept is stated synthetically, but clearly Bohr postulates that the field of force would in a random way either hold back the particles or allow them to escape, in conformity with the Principle of Indifference, and that consequently the probability of escaping would be expressed by the ratio between the number of escaped particles — whatever it may be — in the interval considered and the total number of particles. In short, in the delicate field of quantum physics, the very discerning head of the school of Copenhagen did not hesitate to accept the Principle of Indifference and also the empirical law of chance. The relative frequency of the particles emitted was envisaged as the probability of extracting coloured balls from an unknown urn, or of chancing on stars of a certain magnitude.

In all the above examples the same attitude is met with as is assumed with playing-cards, because the unbiased extraction of a card of any one suit depends on one's having previously considered all the spades, hearts, diamonds and clubs as identical. And if this appears less evident with coloured balls, nevertheless anyone examining them closely would discover differences between those of the same colour, although they are considered identical for the purposes of the game.

The colour of the balls, a card of the spade suit, a star of the fourth magnitude, and the electron which escapes from the field of force of the atomic nucleus are inductions expressed by common nouns; when the facts connected with these linguistic forms become similar to games of chance, they satisfy — whether one willingly accepts it or not — the Principle of Indifference, for the same reason as Buridan's ass died of hunger because unable to choose between two "identical" bales of straw.

4453   The Principle of Indifference and induction

The following experiment is suggested. Books on astronomy supply the classification of the stars, as shown in section 4451. Put into a sack 5719 identical balls, 595 of them white: this will be a model of the sky with its visible stars, in which the stars of the fourth magnitude are distinguished. The most meticulous precautions have been taken that nothing either technical or subjective can affect the draws, which will be carried out in m sets of trials, each composed of a sample of n.

Suppose the theoretical probability P of drawing a white ball is known and equals $595/5719$; then, using the calculus of probability, all the theoretical results of the experiment can be calculated (section 4443) and theoretical limits assigned to the deviations. Undoubtedly, the model constructed is much nearer to the mathematical scheme than it is to the stars in the sky; and in this sense, far from making the experiment "objective" (as it would have been judged in the past) it conceptualizes it and in a certain way places it outside of space and time. A random comparison of a section of the sky could show results agreeing or disagreeing with the draws (theoretical or actual) based on the model.

A discrepancy will probably be found, attributable to the non-random distribution in the sky of stars of varying magnitude, and also, perhaps, subsidiarily, to the inexact enumeration given in books on astronomy — not to mention the possible appearance of some visible star during the period when the experimenter with his model has, so to speak, arrested the flow of time and inductively rendered the future consistent with the past.

But we will not return here to the subjectivity of induction. It is only necessary to point out that all the experimentation described above is based on the Principle of Indifference. The probability of a white ball coming out is identical at every trial; that of identifying a star of the fourth magnitude conforms, or does not, to that principle according to whether the trials agree with the drawing of balls. In conclusion, in the scheme proposed, the Principle of Indifference is not only justified but is shown to be a suitable instrument for verifying induction.

45 Probability and Universal Causation

It is now time to come to the problem of Universal Causation, today generally contrasted with the opposite dialectic of the randomness of natural events.

Practically every book dealing with the Calculus of Probability quotes this passage from the first pages of the *Essai philosophique* of Laplace: "An intelligence which, at a given moment, could know all the physical forces animating Nature . . . would include in one formula the movements of the largest bodies in the universe and those of the lightest atom. For it, nothing would be uncertain: the future and the past would be present to its eyes."

In the preface to the Italian edition of the *Essai*, F. Albergamo says correctly that, according to Laplace, chance is a simple expedient to which men have recourse, being ignorant of the true causes of phenomena. Consequently, probability is not inherent in the universe, as

Buffon thought, but is the "most felicitous way of making up for the ignorance and weakness of the human mind".

Laplace acquired an undisputed position among French mathematicians. Cournot himself, who had a serious spiritual bent and therefore wished, unlike Laplace, to separate determinism from the domain of the spirit, did not hesitate to affirm that the obstacles to foreseeing the future depend on the imperfections of knowledge and the instruments of research.

A hundred years after Laplace, Poincaré wrote, paraphrasing his great antecedent: "An intellect of infinite power, infinitely informed of the laws of Nature, could have foreseen everything from the beginning of the centuries". And in "everything" Poincaré included the organic, the inorganic, and the moral world. "Nous sommes devenus déterministes absolus" were the actual words he used, 150 years after Hume (section 3322); while Mach, Poincaré's contemporary, was contradicting him (section 3111) with the remark: "tout phénomène, si minime qu'il soit, a une cause". I have quoted these words of Poincaré because they express the almost universal mentality of the scientific world up to World War I.

This conviction of the universal concatenation of events from the origin to the end of the world was so rooted in Poincaré that in his work not only does the myth of the "beginning of the centuries" figure, but the eschatological outcome is described. "Dans des milliards de milliards de siècles", he wrote, still evaluating the consequences of expected physical variation of the universe, "sans doute les hommes ne pourront plus vivre et devront faire place à d'autres êtres."

Such a lively faith, in men who no longer countenanced anything transcendental, shows the metaphysical absurdities to which positivism had led men's minds[1].

451   The quantum revolution and probability

In 1912, when the last edition of Poincaré's *Calcul des Probabilités* appeared, the greatest intellectual revolution of modern times was at our doors: the irruption of relativity and quantum physics (section 112) and the philosophical re-thinking of the problems of Nature, originating from the meditations of Russell and Wittgenstein and cul-

---

[1] These and other quotations on this subject from earlier and modern French probabilists can be found in M. Boldrini, "La théorie des probabilities", quoted above.

minating in Neopositivism (sections 14, 141, 143) and the subsequent scientific philosophies.

As already mentioned, the Principle of Universal Causation derives from traditional philosophy, but also, it should be added, from the achievements of classical physics. These achievements have from time to time influenced modern statistical mechanics, which is in fact an application of the calculus of probability, and have endorsed Laplace's concept, namely, the implicit premiss that, because it is impossible in microphysics to particularize the state of a collection of myriads of molecules and to measure the position and velocity of each at every instant, one must resort to the calculus of probability.

But it would be a forced resort. It is true, as Reichenbach points out[1], that Boltzmann's great discovery had shown that the second law of thermodynamics is a probabilistic and not a causal law. However, in classical physics we still continue to admit the deterministic principle according to which, given the state of a physical system isolated at a given instant, the state of the system at any moment before or after this turns out to be uniquely determined by the equations of the appropriate theory[2]. It is scarcely necessary to add — so obvious is it — that experimental observations of the system necessarily exhibit that variability which is ideally eliminated when one has recourse to mathematical devices.

4511 The causal and the casual in physics

Evidently, the revolution represented by the above theories, and by the mathematics on which they are based, would have had not only a scientific but also a philosophic repercussion if there had been inquiry as to why greater authority (other than that of traditional philosophy or of Laplace) should be attributed in science to the principle of causation than to that of probability. What is always and everywhere noticeable in Nature is its variability, never its constancy: a variability which is chaotic only in the single observed element, but regular — and hence describable mathematically — if observed in groups of elements inductively regularized (section 354). Consequently, as scientists wished to remain anchored to determinism and being unable to in-

---

[1] H. Reichenbach, *Philosophic Foundations of Quantum Mechanics*, University of California Press, Berkeley and Los Angeles, 1954, I, 1.

[2] These and other concepts mentioned later are summarized in the study by P. Caldirola and A. Loinger, "L'interpretazione della teoria quantistica", in *La Scuola in Azione*, 1959–60, No. 19.

voke the unity of cases as producing diversity of effects, they had to accept and adapt the theorem of Gauss–Quetelet (sections 381 and 4443). This was supposed to supply a theoretical explanation of the variations in the effects of the same cause, extending it to many diverse cases; while if they had adopted a purely empirical point of view, they would have been able, by interpreting the principle more correctly, to arrive at the hypothesis of the random character of natural facts whose behaviour can be interpreted mathematically when observed in large numbers. This may be really random (meaning empirical) even when, as in macrophysics, a rigorous determinism seems to prevail.

This would have been the state of affairs if, immediately after World War I, the quantum physicists, who seemed to have reached to the very roots of natural facts, had not convulsed the time-honoured philosophic basis of classical physics and its triumphs, associated with the names of Newton, Maxwell and Einstein.

452   Experiments on interference

The facts are well known. A weak current of electrons arrives at a screen in which there are two holes, behind which is placed a detector (for example, a photographic plate)[1]. Owing to the corpuscular nature of electrons, the signals show up as points which determine a probability distribution with several peaks that are called "multimodal" and are characteristic of the phenomena of interference. This distribution leads one to think that the electrons pass through the two holes in an unknown and complicated way. But if the experiment is repeated, closing first one hole and then the other, two probability distributions are obtained, of the type shown in the last column of the table in section 4443 and, more generally, of the Gauss–Quetelet type (section 381). These distributions are called "unimodal" since they have a single peak. The sum of the two distributions is also unimodal[2], and so differs radically from the previous interference curve. All this sounds like a very complicated game of probability. We therefore try to discern the path taken by the electrons passing through the two holes, by means of a light-source situated behind the holes, so that each electron, in passing,

---

[1] P. Caldirola and A. Loinger, *op. cit.*; also H. Reichenbach, *op. cit.*, I, 7, and *The Rise of Scientific Philosophy*, University of California Press, Berkeley and Los Angeles, 1954, pp. 153–4.

[2] It should be noted that similar behaviour has been attributed to the distribution of the inflorescences of the plant order Compositae in section 381, when it was suggested that the sum of the unimodal distributions of the single species would itself be unimodal.

diffuses the light which collides with it. By this device, however, the distribution of the electrons hitting the detector is altered, and becomes of the same type as the two unimodal distributions obtained by opening one hole at a time.

The action of light on the trajectory of the electrons is substantial, say the physicists; it cannot be eliminated and is inherent in the discontinuity of the photons which collide with the electrons, so that control of the passages alters the law of distribution of the electrons.

453   Heisenberg's Uncertainty Principle

On this and analogous experimental data is founded a mathematical theory to which Schrödinger and Heisenberg have contributed. Let q be the position of a material particle, v its velocity and m its mass, and hence also $p = mv$, the momentum[1]. In classical physics, at a given time t the parameters q and p determine exactly the physical state of the particles. Allowing for experimental error, to measure them one must use the Gaussian functions of probability density $d(q)$ and $d(p)$, which are independent of one another. But the independence is less in quantum physics, according to which the two functions at time t can be derived from a single function $\psi(q)$, which has the character of a wave spectrum.

Concerning the variable q, we have

$$|\psi(q)|^2 = d(q). \tag{1}$$

To trace back the link with p, one must remember the character of the wave spectrum of ψ, and consequently we have recourse to the well-known Fourier method of harmonic analysis. The amplitude ϕ of each elementary wave of the spectrum is independent of q but depends on the wavelength λ, for which we write $\phi(\lambda)$. At every wavelength there is an impulse of $p = h/\lambda$, where h is Planck's constant.

It can be shown that the values of $d(p)$ are supplied by the expression

$$d(p) = h^{-3} |\phi(\lambda)|^2, \tag{2}$$

which is called the "law of spectral decomposition".

Since both λ and ϕ are characteristics of ψ — and for this, equation (1) is valid — the law of spectral decomposition leads to the required link between q and p.

Now, the derivation of both functions of the two variables from ψ leads immediately to the Principle of Indeterminacy. By means of

---

[1] We are here briefly summarizing the theory already quoted in Reichenbach's *Philosophic Foundations of Quantum Mechanics*, with convenient adaptation of the symbols.

Fourier analysis, physicists show that the distributions of $d(q)$ and $d(p)$ are such that, calling σ_q and σ_p the respective standard deviations, one arrives at the following inequality:

$$\sigma_q . \sigma_p \geqslant h/2\pi. \tag{3}$$

Since the second member is constant, it can be seen that the more q varies (or the less precise it is), the less variable (that is, more precise) its link p will be, and vice versa. Thus there exists an inverse correlation of the measures of position and impulse, expressible in terms of probability on the basis of the respective values of σ, which is a particular parameter well known to probabilists[1].

A completely parallel relation is valid between the time of observation t and the energy H, and we have

$$\sigma_t . \sigma_H \geqslant h/2\pi. \tag{4}$$

The expressions (3) and (4) are Heisenberg's well-known formulae of Uncertainty or Indeterminacy. They have a general application, although valid only in the subatomic field because of the smallness of h, which is of the order of 10^{-27}. But it should immediately be added (following Braithwaite and Reichenbach) that we are concerned with principles suggested by experience, and that consequently their formulation — which indirectly has an inductive character — has nothing absolute about it. Opposing a new idea to the classical one of Universal Causation, they eliminate from the latter any metaphysical content and further confirm the transitory or historical value of all scientific concepts.

4531   Objections to Heisenberg's principle

It should not be thought that the concepts of quantum theory are universally accepted by physicists. On the contrary, numerous ob-

---

[1] The standard deviation is a measure of dispersion derived from the distribution of the results of draws, which distribution, according to the theorem of Gauss–Quetelet mentioned in section 381, gives the "errors" to which the measures are subject. The formula for the sampling of attributes is $\sigma^2 = rsn$, that is, σ is the square root of the product of the probability r of an event, of the probability s of its contrary $(s = 1 - r)$ and of the number n of trials. σ varies in direct ratio to the magnitude of the "errors" and, for $r = \frac{1}{2}$, depends only on n. A different expression used by probabilists is $k = 1/\{\sigma\sqrt{2}\}$, which is a parameter and can always be substituted for σ since it varies inversely as σ — in particular, increasing as the errors diminish. For this reason, as σ is called the mean square "error", k has been given the name of "precision" of the measures. Substituting k for σ in formula (3), this becomes $k_q . k_p \geqslant \pi/h$, which seems more in agreement with the idea that however much one gains in precision in measuring q, one loses as much in measuring p, given that, as in (3), the right-hand side is a constant.

jections have been brought against them, or at least against certain ways of expressing them, and (says Heisenberg) generally in the name of the concept of reality according to which all constituents of the physical world have an objective existence, that is, independently of whether they are being observed or not. When such names as Planck, Einstein, von Laue and Schrödinger himself, who contributed so much to the new formulation, appear among the objectors, we can understand the great responsibility felt by the creators of subatomic physics. The case of Louis de Broglie is particularly worth noting. He accepted indeterminacy and its implications, at first hesitantly and only later decisively, but in subsequent years reverted to a more positive interpretation of the physical world, proposing to "remove some of the obstacles to the theory" and allowing that quantum physics "described the structure of the various species of corpuscles and forecast their properties, reconciling it with Einstein's field theory in relativity physics".[1]

4532   Epistemological significance of indeterminacy

On the philosophic plane, apart from the mathematical, Braithwaite maintains that at present nothing justifies the thought or the hope that Heisenberg's principle can be eliminated from quantum physics in the near future; and for this reason the most practically acceptable picture of the world is that which considers an element of probability as intimately connected with it[2].

But this is too radical a view, because arguments analogous to those developed concerning classical physics allow every scientific theory — and hence that of indeterminacy — to be considered merely as a transitory and convenient standpoint.

In reality, the position of the indeterminists is logically more valid, because it does not, as does determinism, appeal to the external principles of the Uniformity of Nature and of Universal Causation, but is

---

[1] L. de Broglie, *Nouvelles perspectives en microphysique*, Albin Michel, Paris, 1956. See also an earlier volume by the same author, adhering to indeterminism, the Italian translation of which is *Le rivelazioni della microfisica*, Einaudi, Turin, 1950. A work with an introduction by de Broglie, and with the same deterministic point of view as his, is: David Bohm, *Causality and Chance in Modern Physics*, Routledge and Kegan Paul, London, 3rd edn, 1959.

[2] R. B. Braithwaite, *Scientific Explanation: a Study of the Function of Theory, Probability and Law in Science*, Cambridge University Press, 1955; see also H. Reichenbach, *Philosophic Foundations of Quantum Mechanics*.

G

based on experimental data and mathematical foundations inherent in the problem. But one cannot refuse to recognize that the mathematical theory, despite its rigour and admitted elegance, is, in a certain sense, derived *post hoc* from the experimental data. These latter, however valid they may be, are never absolute and depend either on the underlying information or on instruments.

But these considerations are peripheral to the present subject. What needs to be emphasized here is that, bearing in mind the results of modern subatomic physics and the accent placed by it on the indeterminacy of the phenomena of interference, one is led to look back at the conclusions of classical science and to recognize in it also a predominance of probability. This is contrary to the assumption of a rigorous determinism which is revealed neither by the experimental facts nor by the mathematical formulation and is based essentially on abstract principles external to the actual fields of inquiry.

We conclude that, far from adhering to the axiom of Universal Causation, already rejected by Hume and Mach, modern physicists and the new epistemologists accept the idea that at the root of natural phenomena there is a probabilistic content, which also has an influence on macroscopic phenomena. This is particularly interesting, not only inasmuch as it amounts to the rejection of another among the many axioms on which the earlier science was built—such as the *a priori* character of the basis of mathematics and the principle of the Uniformity of Nature — but also because the substitution of the probable for the determinate tallies better with all the epistemology of the new science, in which objectivity and subjectivity intermingle, so creating links between phenomena, which are modifiable as knowledge and interpretations change (though such links are more likely than certain).

46   The enumeration of classified events is a statistical datum

The reader will remember, as I do, having learnt in early schooldays that only homogeneous quantities can be added or subtracted — a golden rule which we recall even when the conviction that it is completely wrong is long since established. In mathematics the quantities which are added are the numbers of arithmetic and the symbols of algebra and geometry; in each case completely abstract entities, containing in themselves alone the reasons for homogeneity or heterogeneity. Thus the sum of two arcs of the same circle makes sense, but not the sum of an arc and a logarithm. But primary teaching had to do with adding concrete numbers representing things, and this meant that it had to be concerned with adding and subtracting like things.

The foregoing discussion has shown that things do not exist as historically individuated entities which are homogeneous among themselves, but that the homogeneity of objective data is created mentally, is reduced to events or statistical "cases" (sections 3521, 354), and is expressed linguistically by common nouns. Then, and only then, is there meaning in the statement that there are 20 first-magnitude stars in the sky (section 4451), each of which, apart from its specific identifiability, is a "case".

In saying this, one limits the elementary rules of arithmetic to dealing only with concrete homogeneous quantities, but the horizon is greatly widened thereby. That rule — which would be negated whenever one tried to apply it to homogeneous historical data, because the probability of the repetition of an entity is mathematically almost nil (section 4347) — acquires universal application with regard to all the events or cases formulated by man. It is the common noun, the linguistic datum expressing classified events, which makes arithmetical computations possible. And since (as already shown) classifications are extended vertically from classes to classes of classes, and horizontally so that natural facts can be classified according to overlapping criteria, one concludes that the categories of homogeneous elements to which arithmetic applies are extraordinarily numerous, perhaps more numerous than the things themselves.

Thus, we have seen that the stars have been counted according to each category (artificial) of magnitude. Furthermore, when the categories so established are re-classified in the wider group of "stars visible to the naked eye", the six partial computations total to the number 5719, which corresponds to the wonderful sparkle in the sky on a moonless night. It is evident that "visible stars" is a name with a limitation. In a higher class, "stars visible to the naked eye" could be added to "instrumental stars", and a much greater number could be determined for the entire stellar population as known by modern methods.

461 A statistic is a concrete number

There is no need to develop the example further, or indeed to add other different ones. Counting things, phenomena or classified events — that is, cases — is the normal work of life and of science; the laundress does it when writing out the laundry list; the sociologist in making a population census; the philologist when compiling lists of words, phonemes, sounds, variants, foreign words and stylemes in literary works. The astronomer applies himself to enumerating the stars, planets, asteroids and all the bodies wandering in space, while the archaeologist counts vases, and the historian carries out research concerning officials in charge of Roman public works (section 321).

It is an ancient marvel, yet always new, this of "concrete" numbers — the origin, it would seem, of the "abstract numbers" of mathematics (and not vice versa). Without their help, existence itself would be almost impossible. Everything, or nearly everything, in the world is classified and counted. This is why, among the first things learnt, we always remember that man has two eyes, twenty digits, thirty-two teeth and twenty-four ribs; and of things learnt later, that there exist quadruped mammals, diptera with two wings and six feet, and octopodal molluscs; and further, that there are 92 naturally occurring chemical elements, each with a certain internal numerical constitution and a precise number of isotopes.

462 Universality of the concept of a statistic as a concrete number

A concrete number — that is, one containing a plurality of classified elements — is a "statistic" or rather a "statistical fact" in the most elementary meaning of the term. Concrete measurements are also included because they consist of homogeneous elementary units.

Following the nomenclature introduced in section 354 and used since, we recognize that a statistic confers numerical precision on an event: it expresses by a number the classified units or cases which compose the event.

Now, looking back, we are aware of having continually worked with "statistics". The six faces of a die, the 90 balls of a tombola, the 52 cards at poker are statistics, as is the "population" of the stars. And, when counted or measured, electrons, photons, the molecules in a drop of water, the dimensions of camomile flowers and of sunflowers, and every other number or measure of things are also statistics.

The elementary definition of a statistic as a concrete number could be applied (to stretch a point) to the common noun itself which, in relation to aggregates of classified entities, is never universal but only essentially plural. In saying "book", one does not mean "a certain book" but that complex of abstract experiences which the name conjures up. It would not be possible to translate it into figures because it concerns an idea handed down from the time when people first read and wrote. But there are common nouns, such as those of certain very rare illnesses, of archaeological fragments, of certain precious incunabula, of stamps most highly prized by collectors, and of special varieties of orchid, which would remain unexpressive if, in the treatises, catalogues and spoken references, they were not regularly accompanied by the number of units known to exist.

The point should be stressed, because even today it is often said that the name "statistic" should be given only to concrete numbers that express particular categories of phenomena (e.g. social facts, but not facts of natural science) or perhaps to those large counting operations which are statistics by antonomasia, such as censuses and large-scale national inventories[1]. The truth is that no distinction by size makes sense on the conceptual plane, and consequently the inventory of a draper's shop and the inventory of the goods of an entire nation have an equal claim to be called "statistics"!

That the majority of statistics, considered in the widest sense, may end with data-processing or lend themselves to the most laborious and sophisticated research is quite another matter, which has nothing to do with the idea of concrete number.

4621 Concrete numbers are linguistic expressions

If a classified event is a common noun — that is, a linguistic expression — the concrete numbers which describe, in the particular manner of Statistics, the complexes of things and classified phenomena are also linguistic expressions.

On consideration, the idea "a star of the first magnitude" in scientific language is, for the phenomenon in question, a particular case, an elementary entity. The statistical fact "20 stars of the first magnitude" is, on the contrary, a complex proposition because it implies the preceding one, and also the discovery and counting of all the members to which that common noun refers.

Expressed logically, the preceding sentence can in fact be reduced to the propositional function (section 35), "x is a star of the first magnitude". As already stated, it becomes true for all valid substitutions of x, that is, for the individual names of that class of star: Aldebaran, Altair, Betelgeuse, etc. As noted in section 3521, by substituting a proper noun for x the propositional functions still remain in the logical

---

[1] The principal functions used in Statistics, such as means, standard deviation and so on, are now called "statistics" in international literature. But this technical usage is excluded from the present exposition, in which Statistics is presented only in the epistemological framework of the natural sciences. In physics, however, the term "statistics" is applied to certain probabilistic models (e.g., Fermi–Dirac statistics, Bose–Einstein statistics) which concern the theoretical behaviour of subatomic entities subject to random conditions.

field. We move into the scientific field when x is replaced by a species of the kind expressed by the predicate. Hence, in Statistics the statement "20 stars of the first magnitude" is still a propositional function, but it is already determined by the assembly in the number "20" of all the possible valid substitutions for x — that is, in the enunciation of the entire class to which the statement itself refers. We are not dealing here with one of those expressions which Russell calls "extensive", because the list of members of the class is not given. Neither is it an "intensive" expression, because it does not merely explain a property of the class but deals with a new category, empirical and consequently different from the categories of abstract logic, which is designated as "statistical". There are no names of stars given, but it is clearly stated that, once inductively made uniform, they can be counted and the number expressing their "synthetic" list is 20. In the statistical function, identification of the single members of the class is rendered impossible because of the effect of their inductive uniformity. But that function is enough to falsify an "extensive" expression containing more or less than 20 names, and likewise an "intensive" expression such as "the myriads of dazzling stars which not even the full moon can dim".

We should bear in mind that a common noun, with its classificatory and plural meaning, always has a more complex content, whether it is used indeterminately — as when we say "oak" or "lekythos" or "star of the first magnitude" — or whether it enters into a propositional function with an explicit indication of the number of corresponding observations. In the latter case, the noun and the propositional function form a "molecular proposition" in the sense accepted in logic.

These concepts, then, place Statistics on the level of language, and since the language which Statistics uses is the concrete number, it borrows its formal rigour from mathematics, necessarily rejecting symbolic logic from its armoury, just as the pure mathematician does. To the question whether Statistics forms part of formalized knowledge the answer must be affirmative. The truth of this assumption becomes clearer with a deeper knowledge of the more properly technical aspects of Statistics.

4622   Classificatory paradigms and statistical tables

When statistical data refer to complexes of observations, they are ranked in hierarchies and, when necessary, in genera and species, according to the paradigms already mentioned. Returning to the classification (in section 321) of the officials of Roman public works, the authoress of the inquiry, I. Calabi Limentani, not only collected them into categories, but counted them as well. Hence it is possible to include in the table, besides the common noun of each genus or species, the number of elements comprised in it, thereby adding clarity and validity:—

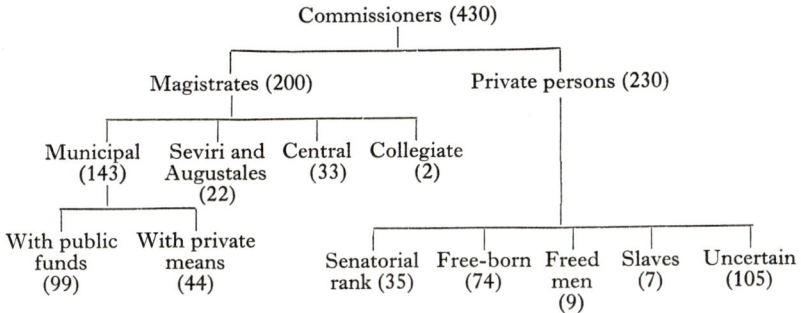

Statisticians, however use a more convenient layout of the facts, arranging them in tabular format in which, for convenience, some classificatory divisions are written at the head of the columns (mostly the genera), while others are placed in the first column corresponding to the rows (species). Compare, for example, the following table, compiled from the facts in the preceding paradigm:—

Subjects entrusted with Roman public works

| | | | Subjects | | |
| | | | Magistrates | | Private | Totals (columns 4+5) |
1	2	3	Totals 4	Private 5	6
2	*Magistrates*				
3	Municipal magistrates:				
4	with public money	99			
5	with private means	44			
6	Total		143		143
7	Leaders of the equestrian order and members of colleges of priests		22		22
8	Central magistrates		33		33
9	Collegiate (*collegi*)		2		2
10	*Private persons*				
11	Of senatorial rank			35	35
12	*Ingenui* and persons with *tria nomina*			74	74
13	Freedmen			9	9
14	Slaves			7	7
15	Unknown			105	105
16	Totals		200	230	430

4623   Enumerable and measurable events and the relevant tables

It was said earlier that classified entities can sometimes be counted (or are "enumerable") and at other times are expressible on a continuous scale by comparing them with a unit of measurement. They are then called "measurable" (sections 353, 382).

In the following table are shown data which are in part computations, in part measures. The table has been compiled from a translation from archaic Greek of a clay tablet belonging to the Mycenaean culture of the fifteenth century B.C., written in "Linear B" and deciphered in recent years[1].

At Sarapeda: contributions for Poseidon and for . . .
(*circa* fifteenth century B.C.)

Villages and other contributors	Commodities in litres				Animals and other goods			
	Grain and flour	Wine	Honey	Unguent	Bulls	Sheep	Cheeses	Skins
Village of Enkheliawon	480	108	6	—	1	—	10	1
Another village	240	72	—	4	—	2	5	1
Military chief	72	24	—	—	—	2	—	—
Religious association	72	12	14	—	—	—	5	—
Totals	864	216	20	4	1	4	20	2

[1] The original of the document here shown in English comes from Pylos, in the south-west Peloponnese (document 171, Py. no. 718). The measures of capacity have been translated by the decipherers into litres, multiplying the original data by 12. The tablet was interpreted thus by Ventris and Chadwick — see M. Ventris and J. Chadwick, *Documents in Mycenaean Greek*, Cambridge University Press, 1956, p. 282; J. Chadwick, *The Decipherment of Linear B*, Cambridge University Press, 1959. For the reduction into litres of the more usual dry and liquid measures as carried out by the two authors, cf. *Documents*, p. 60.

"At Sarapeda (name of the place): Contributions for Poseidon. Contributions for . . .: Enkheliawon (perhaps a village) will give so much: 480 litres of wheat, 108 litres of wine, 1 bull, 10 cheeses, 1 sheepskin, 6 litres of honey. The village (will give) thus: 240 litres of grain, 72 litres of wine, 2 rams, 5 cheeses, 4 litres of unguent, 1 sheepskin. Lawagetas (military chief) will give so much: 2 rams, 72 litres of flour, 24 litres of wine. Thus the region of the confederacy will give for worship: 72 litres of grain, 12 litres of wine, 5 cheeses, 14 litres of honey." Here the statistical data are drawn up either by measure or by counting. The table shows only one of the possible ways of reducing them to tabular form.

4624   Non-numerical statistics

Besides statistics expressed by counting and measurement, we can also imagine non-numerical statistics containing qualitative attributes such as: much and little; large, medium and small; costly and cheap; heavy and light; transitory and durable, and so on. Naturally, specifications such as these give only rough indications; they do not lend themselves to mathematical treatment and are gradually losing their importance in science as numerical expressions replace them. The following table, reproduced by R. Battaglia, shows more than could a long discourse its inadequate information, compared with the numerical evidence of the preceding tables[1]. (The headings *A*, *B*, etc. refer to techniques of production.)

Stone finds in an archaeological excavation of a palae-ethnological site in Gargano, Apulia, Italy

Type of implement	Technique			
	A	*B*	*C*	*D*
Small oval amygdaloids	●			
Hatchets, worked on both faces by chipping	○			
Chisels	■			
Spear-shaped chisels	□			
Spear-shaped objects worked on both faces and oval in cross-section	○			
Spindle-shaped objects		●		
Large knives, triangular in cross-section		○		
Plain knives		■		
Knives for scraping		○		
Simple scrapers		○		
Quadrangular and oval scrapers			○	
Chisels		□	○	
Awls		○	○	
Oval javelin points	□		●	○
Leaf-shaped javelin points				□
Arrows on shafts				●
Stones for throwing	●			

■ Very common   □ Common   ○ Rare   ● Very rare

463   "Statistics" and the science of Statistics

But "statistics" are not Statistics, just as an equality is not mathematics and a proverb is not philosophy. Like these two latter subjects, Statistics, rather than being a collection of numbers, is a body of doctrine with its own constitution and the tasks peculiar to it.

---

[1] M. Boldrini, *Statistica, teoria e metodi*, p. 159, tab. IV, XVI.

G•

Let us return to what was said above (sections 313 and following, and section 37) about the integrating process of scientific inquiry, with its three terms — hypothesis, deduction, and induction. Although, for clearness' sake, the ternary process was explained very briefly and illustrated with examples, we said that it has an abstract character and is not necessarily inherent in this or that particular discipline or doctrine. That the process must in itself be empty of factual elements follows at once for anyone considering the purely rational character of its constituents. Deduction and induction are not peculiarities of one or a few scientific disciplines, but are met with in every branch of the natural sciences, so that they follow common procedures although the circumstances differ.

Although mention was made earlier of hypothesis and deduction, the attentive reader will, however, have noted that they both originate in an axiom or in a hypothesis derived from an inductive process. It was said (section 3221) that in many tests the flash from an electric spark in a mixture of hydrogen and oxygen has always produced water and that one or both of the original gases disappears. Imagine a chemist faced for the first time with this phenomenon (as happened to Galileo with the pendulum). One would expect that, to convince himself, he would repeat it over and over again with variations, and take notes; the number of favourable results and the possible exceptions would be accurately recorded. These notes and summarizing calculations would comprise the "statistics" of the imaginary experimenter.

We possess many sets of original notes from ancient and modern masters. There are many hundreds of tests, carefully recorded, which enabled Spallanzani to arrive at the first decisive denial — later confirmed by Pasteur — of spontaneous generation. The original notes are also known of the long and complicated research by means of which the Abbé Mendel was enabled to formulate the principles of heredity, and those which led Marie Curie to the discovery of polonium and radium[1]. It will be recalled that "protocol propositions" are the first verbal notes by which the natural scientist records his daily observations, and "molecular propositions" are the results which can be enumerated and expressed by common nouns (section 4621).

---

[1] For Spallanzani's statistics see M. Boldrini, "Lazzaro Spallanzani", in *Annuario della Scuola de Studi Superiori sugli Idrocarburi*, 1958–9, p. 51. For the question of the priority of Spallanzani over Pasteur as regards protozoa, see M. Boldrini, "Louis Pasteur", in *La Scuola in Azione*, 1959–60, No. 6. For one of the original sets of notes of M. Curie and for an analysis of those of the Abbé Mendel see M. Boldrini, *Statistica*, pp. 112 and 94.

The natural scientist obviously makes the best use of his expertise and of the means to hand when working on statistical documentation. Spallanzani used putrescent liquids for growing his cultures of the "little animals" (protozoa) with which he was experimenting; Mendel crossed several varieties of peas in the garden of his Moravian monastery; and Marie Curie carried out her electrochemical analyses with electric batteries — a multitude of researches and different expedients which have nothing in common beyond the skill, acumen, and infinite patience of those who carried them out.

It needs genius to know when to pass from the particular to the general and to draw from the experiments a hypothesis for research. But at this point rules have to be introduced, because intuition never means the vagaries of genius but order, discipline and method.

What is important to note at the moment is that induction commences with the formation and collection of statistics, that is, with documents on which talent can construct provisional theories in the form of hypotheses, which the expert can then develop deductively in order to confirm or reject them, as the case may be.

464   Induction and deduction in Statistics

It is the concern of the whole deductive development to keep up with new propositions, to be checked for experimental confirmation of the hypothesis. Once again, the intervention of technicians qualified in the particular subjects will be of value as indicating the direction for the deductions, given that chemistry, biology or philology obviously do not follow the same paths.

But here also the procedure is subject to rule, and in documentation or in the more advanced phases of research the method is developed with the aid of mathematical tools. It is the "concrete" mathematics dealt with in Chapter 1 that is in question here; it is less the concern of the pure mathematician (or not his concern as such) than it is of that particular mathematician who is a statistician. It is in fact Statistics which pre-arranges the abstract schemes by means of which concrete quantities and their inter-relationships can be investigated, in order to check hypotheses and form conclusions. And among such tools is included the calculus of probability, discussed in the first part of this chapter: indeed probabilists are sometimes classified among statisticians in general. These remarks show the broad scope of Statistics, which is not confined to the interpretation of concrete numbers, as one might at first think. Statistics is only one of the poles of the process on which scientific induction and deduction are built.

This is the substantial difference between the abstract sciences — mathematics, logic, etc. — and the concrete ones — physics, biology, political economy, etc. One group moves from axiomatic premisses which develop tautologically, while the other starts from experience, and after an inductive process, at the beginning of which are the con- concrete numbers of statistics, returns deductively to experience. It is of such an inductive–deductive process, which the experimenter scrutinizes and carefully checks with the proper instruments, that Statistics consists. This point will be more fully treated later.

47   Probability and Statistics; sex-ratio of births

It now remains to link up more clearly the ideas of probability and Statistics, even if the relationship may seem intuitive.

In the following table are given the number of male births and the total number in Italy in 1958, in a commune, a province, a region, and the whole remaining part of Italy. The probability of a male birth is taken as the ratio between the number of males and the total number of births in the relevant area.

	Males born	Total born	Ratio, males : number born	Relative deviation in comparison with the rest of Italy
Commune of Matelica	54	99	0·545	0·0319
Province of Lecce	6,972	13,724	0·508	0·0046
Veneto, Friuli, Venezia Giulia ..	40,976	80,276	0·510	0·0031
Italy (excluding the previous zones)	398,677	776,379	0·514	—

Assuming such a probability, the relative deviation of the first row becomes:

$$\frac{54}{99} - \frac{398,677}{776,379} = 0\cdot0319.$$

Proceeding thus, we obtain the results of the last column, which establish the decrease of the deviation as the number of observations increases. This points to agreement with Bernoulli's theorem and tallies with the results of section 4442. The latter case seems to differ from the demographic table given above because, since the probability was not known, it was estimated from a large experimental complex, relying on the empirical law of chance. In fact, however, the difference is only apparent. In drawing balls from a bag or numbers in a lottery, one adopts as probability the proportions of the colours or numbers and hence one is making an arbitrary hypothesis, replacing ignorance of the true probabilities by estimates based on these particular instruments of gaming.

471   Mendelian experiments with peas

The well-known experiments in crossing peas, which made the Abbé Mendel famous and were the springboard for the science of genetics, are based on principles which affirm that from the second generation onward the probability of a yellow pea appearing in a yellow–green cross is 0·75.

Here are some experimental results:

	Yellow seeds	Total number of seeds	Ratio	Relative deviation
Experiments by Tschermak	3,000	3,959	0·758	0·008
Experiments by various authors ..	152,824	203,500	0·751	0·001

The deviation from 0·75 is in accordance with the theoretical considerations mentioned above.

4711   The subjective element in statistical experiments

From the epistemological viewpoint, the examples of the two preceding sections must be evaluated for what they are and not for what they are generally believed to be — i.e. not as objective descriptions of natural facts but as arrangements and scientific interpretations (that is, inductive in the sense explained in Chapter 3) of classified entities. We mention only in passing that the notion of a sex-ratio involves reducing one of the most complex phenomena of human life to the level of drawing black or white balls. The example of the peas is easier to analyse. The descendants of peas of a yellow-by-green cross present a continuous scale of colours from yellow to green. The distribution of the two opposite allelomorphs in the Mendelian sense is undoubtedly somewhat arbitrary, and it can be seen at first glance, or based on the well-founded hypothesis mentioned in section 471, that the yellow descendants are about 75 per cent of the total. Thus, setting out 400 peas and grading the colour by sight, the observer would judge approximately the first 300 to be yellow and the remainder green. Naturally, from one experiment to another the percentage would vary according

to Bernoulli's theorem. On the other hand, it has been shown that the
variations entered up in Mendel's notes and in those of the first naturalists who, after the rediscovery of the famous laws of heredity at the
beginning of this century, repeated his experiments, are less than the
theoretical expectation — a clear sign that the anticipated 75 per cent
considerably influenced the classifying of hybrid peas between the two
colours. One is faced here with a manifestation (actually verging on
prejudice) of that adaptation of facts to theory which was discussed at
length in Chapter 3 (cf. particularly section 364). It constitutes one of
the logical ingredients of induction, which were unsuspected at one
time but are now well understood[1].

472   Experimental data and the theory of probability

We have already seen (section 4453) that the possible disagreement
of an experimental result with the theoretical expectation in games of
chance can, paradoxically enough, constitute a confirmation of the
sound basis of the principles of the calculus of probability.

The statistical example examined in section 47 showed that in
four areas in Italy more males than females were born in 1958. The
constant excess of masculine births is a phenomenon which occurs in
all places and at all times with astonishing regularity, and has been
the subject of lengthy discussion among demographers since it was
discovered in the mid-seventeenth century. It was thought that the
phenomenon was ruled by simple probability. If 100 balls are put in a
bag, 51 of them white, in the long run the draws (the balls being always
replaced) tend to give precisely that proportion of about 0·51 which
Statistics indicates. As the births in successive years can be considered
as separate series of draws, the slight annual variations of the male
ratio appeared similar to the well-known oscillations in the results of
several series of draws, like those considered in section 4443. But the
celebrated German statistician Wilhelm Lexis, at the end of the nineteenth century, studied many series of sex-ratios and stated that these
did not vary as forecasts should which are based on Bernoulli's theorem.
In a new mathematical scheme he was able to show that the probability
of masculine births varies with time and place. It was as if the composition of the balls in the symbolic bag from which the new-born were
extracted was from time to time slightly modified. The word "slightly"
is here used to denote the persistence, in bringing about this pheno-

---

[1] For this demonstration see M. Boldrini, "Sulla dispersione dei caratteri
mendeliani", in *Acta* of the Pontificia Academia Scientiarum, 1941.

menon, of constant factors which would, however, be accompanied by secondary changing circumstances.

Demographers think that they have only in part identified these circumstances, and they will not be considered here.

This example serves to bring out the practical importance of the points already made about assuming equal possibilities on the basis of which the probability is calculated. In the imaginary game of drawing out a male or a female, according to the findings and interpretation of Lexis, the chances would not continue to be equally possible over a long period of observation.

But the evidence of the births demonstrates the part played by the three-fold process of scientific inquiry. Statistical observation has shown a relatively constant excess of male births. The hypothesis was advanced that the phenomenon was similar to an ordinary lottery. Deductive development of the hypothesis and further experimental comparison proved that this was not true. Lexis had to put forward and test a new probabilistic scheme, in which the facts agreed with the theory, so that a satisfactory explanation would be forthcoming. The merit of Lexis — praised by Keynes and inexplicably belittled by Borel — has been confirmed in modern developments of statistical mathematics.

473   Statistics, probability and induction in relation to finite populations

The ideas introduced above can be made clearer when one keeps in mind the relation of Statistics with the subjects of probability and induction.

Proceeding systematically, we shall first examine the case of those statistics which are concerned with finite complexes and so include, or could theoretically include, all possible cases of the event to which they refer.

Statisticians have given the name "universe" to the totality of existing individuals. (The more modern English term in Statistics for "universe" is "population".) A subset drawn from the population by some rule of procedure is called a "sample".

As stated above, the statistics of finite populations can include, at least theoretically, the totality of experience: this is the case with the most elementary "statistics" such as the laundry list or the catalogue of a private library and also for those other scientific enumerations whose great size is insufficient to differentiate them from the former. Such could be the counting of all the visible stars in the firmament, the classification of all the cells of an elementary organism, the collecting of the statements of age of the entire population of a country, the recording by inventory or book-keeping of the goods of a business concern (as is in fact the case with the contributions to the worship of Poseidon in the fifteenth century B.C.—see the table in section 4623).

Statistics of this and of a similar type are exhaustive and complete, and they cover a universal experience: that is, they supply a quantitative description of the entire population. Furthermore, as they express a definite situation of fact, they have an eminently "historical" characterization (in a loose sense), though they are not always interpretable without ambiguity. This agrees with the thesis, repeatedly presented above, that science is concerned not with objective realities but only with linguistic expressions of human experience. Thus the stars whose statistics recur in this book (see section 4451) are those "declared visible by the astronomer Houzeau". Their number, 5,719, differs by much or by little from the figures given by other authors. Among these should be mentioned Proctor, who gives the number as 5,850, adding that Heis and Gould, with their very acute vision, counted up to 10–12 thousand (including, however, those of the seventh magnitude). This means that certain stars are included, or not, in the category "visible" according to the author one happens to quote.

Analogously, population censuses mostly give — as the total of the entire population—at least two numbers for any given time, almost as if the same people could exist and not exist in the same region. But this is because the "historical" interpretation of the population can vary (e.g. by including, or not, numbers of people occasionally present in the country, foreigners, etc.).

4731   Samples from finite populations

But groups which are finite and hence, in theory, always enumerable and measurable as a population are not always perceived as a whole. And then, to the inherent difficulties of data interpretation — as we have already affirmed regarding universal situations — are added others which at times perplex, and at times actually lead astray, someone wishing to deduce more than is immediately apparent[1].

The result of market research intended to find out, by a small sample, whether some projected merchandise will be well received by at least 10 per cent of its potential consumers could be indecisive. This is

---

[1] For a systematic treatment, see M. Boldrini and A. Naddeo, *Le statistiche empiriche e la teoria dei campioni*, Giuffrè, Milan, 1957.

because a sample of the finite population of these consumers does not necessarily represent them with complete accuracy.

But here we are dealing with the easiest case, because it is subject to a simple doubt. Consider, on the other hand, the table in section 4622, giving the classification of Roman public works according to the qualifications of those concerned, and other specific information. The information regarding the works in question is complete in itself, but it comprises a quite special sample of a population, certainly limited, but concerning which a great many relevant facts have been lost. Such a sample would furnish historically valid data — that is, it would express the relative frequency with which, for instance, the municipal magistrates were in the habit of dedicating public works erected either with public funds or at their own expense — only if it constituted a miniature but faithful picture of the original population of which it is now only a remnant. That it in fact does so must be denied, because of the destruction which through the centuries has overwhelmed the cities and provinces of the Roman Empire.

4732 Bernoullian sampling, the principle of chance, and induction

In general terms, the questions implicit in the preceding remarks become clear when put in the framework of the principle of chance (section 4342).

Let $x_1 + x_2 + \ldots = N$ be the frequencies deriving from the modality of a phenomenon in a finite population x of size N.

If they are all reduced by a constant factor k, the proportional frequencies remain unchanged because the expression $k(x_1 + x_2 + \ldots) = kN$ is identical with the preceding one. But if the reduction was made by drawing a random sample of the population, by the principle of chance the coefficients would be k_1, k_2, $\ldots$, oscillating at random round k, and we should have $k_1 x_1 + k_2 x_2 + \ldots \sim kN$, with the presumption that the expression would tend to equality as the absolute frequencies increased. A sample drawn by chance from this population is called a "random sample" or a "Bernoullian sample". The law of chance also states, however, that for non-random samples the preceding expression only exceptionally becomes an equality. Non-random samples are said by statisticians to have "bias". The principle of chance has, as we have seen, an essentially pragmatic or, so to say, historical validity. But when inserted into the theory of induction it obviously becomes applicable to the future. On this point, it should be remembered (section 3561) that induction is not a synthetic operation but an analytic one; it is a development of premises, a tautology, and, as such, is a rigorous process. Naturally, the facts could give the lie to inductive forecasts, but this does not invalidate the interpretation: it makes it necessary, however, to modify the premises from which the induction, shown to be false, has been developed.

This is true in general and true likewise regarding the law of chance, whose validity as an instrument for forecasting can be affirmed, inasmuch as from random samples of a finite population, measures can be deduced which can be considered axiomatically as expressions of the *a priori* probabilities inherent in the population itself.

4733 Bernoullian samples and biased samples from finite populations

It was stressed above that the Roman public works, the titles of whose dedicators are known, do not constitute a random sample derived from the original population but are a biased sample. The same negative judgment would seem to apply with regard to the *lekythoi*, the classification of which was presented in section 321 by reference to their design; considering that most of the examples in the museum in Athens came from the famous archaeological cemetery of Eridanos in that city, we can reject the hypothesis that the samples are in general representative.

The above two cases of bias are very evident and would scarcely mislead an expert observer. But it is necessary to be on one's guard in tricky situations, so as not to add errors of judgment to the risks of induction.

Nevertheless, it should be said at once that in Statistics random sample groups are the rule and not the exception. It is precisely from this that the points developed and the theories built up on them by statisticians draw their validity. For example, it is permissible to think that the flowers in Dutch gardens, on a given day, are represented by the variety offered for sale in the great wholesale market in Amsterdam; and that the fish which used to be sold daily in the market of Fiume — from the statistics of which, collected by U. d'Ancona, sprang the famous mathematical inquiries of Vito Volterra on biological association — was a random sample of the species which had found their way during the preceding night into the fisherman's nets in the Gulf of Quarnaro.

In developing countries the major part of economic and demographic inquiries are carried out on random samples of finite populations, since the necessary organization for a complete survey is often lacking. Developing this procedure systematically, Professor Mahalanobis has promoted Indian statistics, and has drawn up an economic plan. Market researches and public opinion polls (Gallup, Doxa, etc.) are also carried out as far as possible on Bernoullian samples of the original limited populations.

474   Bernoullian samples from infinite populations

Without doubt the questions and situations just presented are logically and technically complex. But they do not correspond to the most difficult paradigm or to the most frequent one.

When the samples come from finite populations, it is possible (at least theoretically) to verify their reliability directly. To know whether the marks gained by ten students taken at random from a class are representative of the whole is a question which can easily be answered by comparing with the totality of marks. But what is to be said of infinite populations about which, by definition, we can know nothing? A pendulum could oscillate until the end of time, but we have to trust to samples of measurements of its periodicity in order to formulate and accept the law of Galileo. To return to the example in section 471, it was only from samples of the infinite number of possible crosses between yellow and green peas that the laws of Mendel were evolved. Are they true in general? It is easy to see that on all sides men formulate general rules, starting from limited observations — however numerous they become eventually — derived from infinite populations.

Now the only thing which can be said about these populations is that they are subject to Cantor's paradox (section 353), because, in them, single characteristics ("modalities") can have the same order of infinity as the whole set. For example, the pomaceous trees (pear, apple, hawthorn, etc.) produce flowers with five petals, but a minority with more or fewer petals can be found. But in the infinity of flowers which Nature has so far produced, and will continue to produce, there are neither majorities nor minorities, and it is therefore a question of populations about the measure of whose composition nothing certain can be said. However, one can gather general numerical information about them by relying on random sampling; that is, by the so-called Bernoullian method, and in such a way the unlimited becomes limited.

The probabilities connected with such samples can be justified by the law of chance only by analogy with samples derived from finite populations, but there is the added difficulty that, for these latter, only a control by further samples could be considered, while the idea of increasing their dimensions to the point of containing the entire population would be self-contradictory. In this case the numerical increase of each sample and comparison with others offer the only conventional criteria of the validity of the inductions which are derived from them. This is an interpretation that is open to lengthy discussion, because it links up with the idea of the Kollektiv (section 4341) and likewise with the earlier interpretations of more or less regular phenomena (described by the term, now no longer used, as "typical" or "atypical") which postulated a more or less rapid convergence towards stability as the number of observations increased. But convergence in the von Mises sense is a postulate which, in the mathematical field to which it belongs, remains justified by its deductive fruitfulness, while the experimental convergence of numerically increasing samples drawn from infinite populations is based exclusively on the analogy with finite populations and, naturally, on the validity of the assumed inductive position.

4741   Sampling from infinite populations and induction

In spite of these clearly serious limitations it can be said that all the sciences, from the physical to the biological and social, build up their results by induction from samples of infinite populations about which one actually knows nothing. Stuart Mill's convenient loop-hole (section 333) makes an indirect appeal to the axiom of the Uniformity of Nature to which contemporary science denies the credit formerly accorded to it.

Today it must be said, not that samples reflect the characteristics of the respective infinite populations, but that the characteristics of these latter are interpreted in the light of knowledge supplied by the samples. It is an arbitrary process, therefore, and a contradictory one, though useful, because once the *a priori* probabilities are assumed from the axiomatic premises, they become the legitimate instrument of control of the experimental results. The tautology of the procedure is evident once again, and its legitimization on the experimental plane remains, as always, dependent on inductive confirmation. If such confirmation fails, the whole procedure is useless and must be reconstructed on a different basis.

It must be admitted that nowadays classificatory systems and inductive generalities are stated more and more efficiently in those systems of general ideas which are the pride of modern science. Naturally, those inductive generalizations the usefulness and coherence of which is recognized have become part of the corpus of scientific doctrines, while many other unsuitable, useless, or uncertain ones have been, and are continually being, rejected. Not to set one's sights too high, consider the immense field of science, so different now from what it was forty years ago, and the mass of published work which, with rare exceptions, adds little or nothing to knowledge. A triumph, yes, but also a somewhat formidable picture of a mass of research and of printed matter, not to say of reading — a picture which may escape the inattentive eye, dazzled by the occasional highlights which guide and propel the progress of science.

48   Whether Statistics can be distinguished from Logic

Before concluding this discussion it is opportune to mention a question which preoccupied earlier statisticians and which has cropped up again from time to time. Is it possible, it was asked, to distinguish Statistics from Logic, or should the former be considered as an aspect or at least a branch of the latter?

The question may seem relevant after what has been expounded in this study, because (as we have seen) one comes to the idea of Statistics only after reducing science to language, mastering the ideas of deduction, induction and probability, and discussing many other subjects which are tackled in books of logic. But this is only true because of the simple fact that no strict antithesis could exist between the two disciplines of Statistics and Logic, which have so many points of contact with one another, just as each has with mathematics.

The difference between Statistics and Logic becomes clear when one considers their differing outlooks rather than their common origins.

A comparative development of the ideas of Logic and Statistics, pointing out the similarities and differences, would be out of place here; the interested reader has the means, with today's abundant literature, to be well informed on the problems of this subject, evolved from the ideas to which later commentators, imitating the spirit first of Aristotle and then of Bacon, had given the name *organon* — that is to say, "instrument". And this was not a matter of chance, because in classical Greece, when the technique of reasoning was being elaborated with much insight but before the advent of logistics, Logic was for Aristotle and his successors the theory and technique of syllogistic reasoning (cf. section 161). But the differentiation of the sciences which rules today, and the great modern interest in the problems of Logic — after a long period of indifference — enable the subject to be seen in a very different light from that of antiquity. Logic is no longer only syllogisms; but neither is it, as was long maintained, the science of the laws of thought, because under such a title it would come to be confused with psychology, which itself has long since become autonomous[1].

It is true that to reason is to think, but to think is not always to reason (as, for example, in the party game which starts with the command: Think of a number)[2]. And logic is no longer even an *organon*, in the sense that it does not include among its tasks an inquiry into the obscure ways by which reasoning disentangles itself (this, too, is the job of psychology). Logic has as its only attribute the critical analysis of the validity of the process of reasoning, posing the problem whether a conclusion reached follows from the premises without error, contradiction, or insufficiency.

The distinction between correct and incorrect reasoning is the central problem of logic, and this without any interference with the factual content: there are true "propositions" and false "propositions", there are correct or valid "arguments" and incorrect or invalid "arguments": but whether the factual content — that is, the propositional statement — is true or false does not concern logic at all, provided that the argument is developed in a formally unexceptionable way. In section 1631 we gave some examples which illustrate this.

481 Similarities and differences

The ideas so far developed and their links with Statistics become clear when we consider the following points:—

(1) There exist declaratory expressions, such as "parallel lines meet at infinity" and "Socrates is mortal"; for these the name "proposition" is used. There are propositions pertinent to the abstract disciplines, and these, according to the axiom of Wittgenstein, are empty of content; there are also propositions pertinent to the concrete disciplines, having for that reason a true or false content.

(2) To carry out its main task, Logic has worked out methods and techniques of formulating and of verifying proofs, but not of ascertaining the truth of propositions, this being the concern of the particular disciplines (section 1631).

(3) Statistics joins concrete proofs formally into scientific "structures". Like Logic, Statistics also has a "formal" character, that is to say, the populations under discussion contain fixed parts and variable

---

[1] A clear exposition, from which these ideas are freely developed, can be found in I. M. Copi, *Introduction to Logic*, Macmillan, New York, 1957.

[2] And hence it is not even always an assertion that a discourse is true or false. Aristotle instanced prayer (*é euché*) as an example of non-assertive communication. Cf. H. Scholz, *Storia della logica*, Silva, Milan, 1962, p. 29, note.

parts. It embodies procedures which, where they substitute expressions of content or fact for the variable parts, become complexes of proofs — called "structures" — which are interpretable as true or false.

"Statistical forms" are normally presented as expressions with "variables" (properly so-called) whose numerical values are obtained by experience, and with calculated "parameters" which determine the functional relations between the variables themselves. Statistical forms, however, are not generally limited to functional expressions, but are mostly accompanied by rules of procedure which are part of the entire ternary process "induction–hypothesis–deduction" (which was discussed in Chapter 3, and we shall return to it in Chapter 5). Obviously this does not diminish the formal character of Statistics, which remains essentially external to the specific subject-matter of the scientific disciplines, even where it co-operates closely in formulating their particular problems.

Given, then, both its formal position and its operational task, one could say that, with respect to particular sciences, Statistics finds itself in the same independent position which the "syntactic" rules of logic have with respect to the "semantic" content of the propositions to which they are applied (section 144).

482 Scholz on epistemology and logic

The word "epistemology", as a theory of sciences, was introduced by Aristotle[1] and has passed into current usage. It is employed by Scholz as a generic term to mean formal logic in the deductive sense of Aristotle, or Kantian transcendental logic, or, again, non-formal logic — the logic of judgement. Scholtz writes as follows:—

"(a) Epistemology is a generic concept, comprehending formal and non-formal logic, and may be defined as the theory concerned with the instruments necessary for scientific knowledge in its widest sense.

(b) *Formal* logic coincides, then, with that part of epistemology which formulates the inferential rules necessary for the construction of any science whatever, and which obviously supplies everything required for an exact formulation of these rules.

(c) By *non-formal* logic we understand all the remaining field of this epistemology, that is, on the one hand, what can be subsumed to the concept of this epistemology, and on the other, what is different from formal logic."

---

[1] H. Scholz, *Storia della Logica*, p. 32, note. The Aristotelian text in Greek is given there.

The same author comments in a note a little further on that, at a first approximation, non-formal logic includes the "logic of the natural sciences" (Mach, Duhem, Weyl, Reichenbach, etc.) and the "logic of the sciences of the mind" (Rickert, Spranger, etc.)[1].

It remains to be seen what may be subsumable to the concept of epistemology thus universalized, since "all the rest" — coming after "the inferential rules necessary for the construction of any science whatever" and "everything required for an exact formulation of these rules" — has a dimension even broader than science itself. There would be no harm done if one described as "epistemology" not only formal logic but also this very large remainder, but one would not then attain the clarity sought for in section 481. It is better, therefore, to attribute to the Aristotelian word the plain meaning of "theory of the sciences"; that is to say, on the basis of a nomenclature previously adopted (section 144), to allow it the character of a "metascience", recognizing that one is dealing with a discourse about the sciences and is safeguarding the autonomy of the sciences themselves, each one of which produces and uses its own special instruments of investigation. The word "epistemology" has been used in this meaning and will continue to be so used in this book.

This position again justifies the affirmative character of Logic as the science of formal "subjects"; and beside it is placed Statistics — also a formal science, but concerned with the "structures" (methodological and technical–mathematical) of the natural sciences (*see also* section 172).

49   Outline of a definition of Statistics

A definition of Statistics, taking in the concepts we have been precisely defining and their reduction to a concise formula, appears neither easy nor necessary, and perhaps not even useful. We recall that the various branches of knowledge can be reduced to the two classes of the formal disciplines, which develop deductively postulates proposed *a priori*, and the applied disciplines. A foremost group of these latter derive their axiomatic positions from experience, and from these positions the respective doctrinal systems are then deductively developed and experimentally checked.

Since they emanate from observation of fact, this foremost group of applied disciplines and their particular doctrines merit the name of "natural sciences" — a name which in everyday language comes to be

---

[1] H. Scholz, *op. cit.*, p. 46.

applied restrictively to some of them (section 58 *et seq.*). Every natural science is specialized in the sense that it possesses a factual sector or particular body of doctrine, makes use of its own devices and instruments, and requires a specific education and suitable training on the part of its exponents.

Furthermore, the natural sciences do not present themselves as one block but consist of ideas and doctrines with their own specific development. It is in the intimate constitution of these more specialized branches that one best sees in operation the ternary process of induction–hypothesis–deduction (however the triad may be ordered) which we have been discussing. With the help of this triad, the worker observes the facts of the world on the basis of selected objective externals (characters); lists, enumerates, and measures them; and selects the suitable mathematical tools — often based on the calculus of probabilities — by means of which he prepares the quantitative documentation. In this way the results are obtained and interpreted.

Statistics consists of this procedure and the relevant methods; it is a kind of formal combination of experimentation (understood in the widest sense) with precepts and a technique which, at least in theory, is not concerned with the factual material of individual branches of research, and so constitutes a subject in itself. In such terms we can epitomize the essential characteristics and operational tasks of the discipline we call Statistics — so qualifying it to be the methodological structure of the natural sciences.

5

The methodological structure of the natural sciences

51 Syntactic and semantic analysis of theories

From the practical identification of science with its language (Chapter 1) one can deduce that every theory is formed by a discourse which, in the particular case of the natural sciences, co-ordinates and explains a field of events of the world.

It is possible to analyse the propositions of science according to two different directives, one of which considers the "syntactic" aspect, i.e. makes clear the individual parts of the discourse, explains their connections, and interprets the co-ordinating logic of the whole. The other examines the "semantic" aspect, i.e. refers to the factual content (section 144). Such an analysis of a theory from the double viewpoint of form and fact is necessarily realized in propositional terms, and the whole system will be called "metatheory", in keeping with the concept (already defined) of "metalanguage", which refers to a discourse about an object language, and of "metamathematics", which is concerned with discussion of the doctrines of mathematics (sections 144, 24 and 482). A metatheory is therefore the linguistic analysis of a theory, carried out by considering its propositional structure or by critically evaluating its factual complex.

511 Statistics and Metastatistics

The building-up of a scientific doctrine is, one might say, the "science that one does", and its metatheory is the critical analysis of the "science done". The main task of this book, devoted as it is to the theory of Statistics, is not indeed to teach the procedures used in the science that one "does" — a task better left to the manuals of statistical technique — but to delineate by means of "Metastatistics" the formal syntactic structure of the scientific theories of which Statistics is the fundamental constituent (see Chapter 4).

512   The distinction between logic and psychology

The syntactic character of the proposed metastatistical analysis should be emphasized, and as opportunity arises this will be made clear, so as to resolve the discussions between logicians and psychologists about the limits of their respective competencies.

Piaget states that it is an error of the followers of what is known as *Denkpsychologie* (from Marbe to Selz) to have concentrated on investigating adult intelligence by the method of "stimulated introspection". In this system the structure of the mind comprises complexes which are already stabilized and equilibrated; but it loses sight of their dynamic genesis, which can be perceived only in a child during its intellectual development — i.e. it exhibits psychic activities as fully matured.

In fact, according to Selz, the solution of any problem involves in the first place an "anticipatory plan" which links the end to be attained with a complex of ideas, leaving a lacuna between. Secondly, this plan is "completed" by means of concepts and relations which supplement the "complex" according to the laws of logic[1]. Here it is evidently a question of problems of the abstract disciplines. We need only substitute for the words "anticipatory plan", "completion" and "complex" the usual ones of "axiom" (or hypothesis), "deduction" (or the deductive and inductive process), and "conclusion", (or, as we shall say, "law") to find ourselves fully conversant with the structure of Statistics.

Speaking as a psychologist and not as a statistician, Piaget objects that logic as objectivized by Selz is repugnant to the psychology of the intelligence; it stands to this as strategy and poetics stand to war and poetry: it is static and atomistic. According to Piaget, it is necessary to subjectivize logic, to make it the mirror of thought, to transmute the system of formal logic into the final outlet of those developmental processes which determine adult thinking.

These are problems which it seemed desirable to mention but not to dwell upon, either because the authors quoted above fully justify their positions in their own writings, or because the metatheoretical considerations on the structures of the factual disciplines which I shall give below are not intended to go beyond the syntactic terrain. Even less are they meant to concern themselves with the genesis of ideas and of

---

[1] J. Piaget, *Psicologia dell'intelligenza*, Editrice Universitaria, Florence, 1952, p. 51.

scientific knowledge. In short, we remain on the formal plane (as Selz does) rather than switching to the developmental one (as Piaget thinks, perhaps rightly, is appropriate for psychology).

513 Statistics faced with the variety of scientific disciplines

The greatest difficulty which faces us in syntactic analysis stems from the almost unlimited variety of disciplines and scientific doctrines which envelop the techniques of observation and experimentation and cannot fail to influence rational interpretation and exposition. That this is not an impediment to general analysis is, however, shown by the fact that the sciences, with their very wide range of application, always rest essentially on the common basis of classificatory induction. Moreover the threefold phases of research — hypothesis, deduction, and induction — are of a formal nature (sections 322 and 37), and their universal character is not lessened even when one gets down (as we shall shortly try to do) to a detailed consideration.

52 Axioms as a beginning

It was said in Chapters 2 and 3 that the old conception of axioms as binding, as absolute and evident truths, has given place since the beginning of the present century to a humbler but more workable idea, according to which an axiom would be any affirmation which initiates the deductive formulation and development of a scientific theory.

A modern expert in logic says that every axiom must today be considered as provisional and subject to change, no longer as a "principle" but as a "beginning". He clarifies as follows: "The difference between the two points of view is this: to conceive of the axiom as 'principle' is equivalent to attributing to it, besides the formal characteristic of constituting the starting-point of the deductive chain, the onus of 'justifying' the truth of the deduced propositions. This can only happen if it is a certainty in itself; to conceive of the axiom as simple 'beginning' is equivalent, on the other hand, to leaving out of consideration this last viewpoint."[1]

These valid considerations can be seen to be at work in subjects already discussed. The notions of "zero", "number" and "successor" in Peano's arithmetic are undoubtedly simple affirmations, neither evident nor demonstrable (section 15). Such also are Jeffreys' postulates

---

[1] E. Agazzi, *Introduzione ai problemi dell'assiomatica*, Vita e Pensiero, Milan, 1961, p. 14.

of equiprobability, transitivity and equiconfirmation, put forward in section 4321. They are purely conventional, but indispensable to initiating the chain of deductive developments based on them.

The same applies to the fundamentals of the factual disciplines, which have become purely definitive and conventional after rejecting every principle assumed from outside, such as regularity, simplicity, and the causality of Nature. As has already been recognized, we must attribute the character of purely axiomatic assumptions to the Newtonian concepts of mass and force, to the postulate of the constancy of the velocity of light (section 224), and to the idea of electromagnetic fields of force, as to many other things whose validity derives from the consistency recognized up to the present in the scientific systems built upon them. If and when contradictions emerge, as in the case of stoichiometry (section 3553), contemporary science would not hesitate to correct these concepts, or even consign them to the rubbish heap of useless instruments which has already received pure space, absolute time, the ether, phlogiston, and so forth[1].

521   Axioms and theorems

Just as logicians have demonstrated that axioms and definitions are the same thing (section 22), it can also be shown that theorems and

---

[1] From the study of personality, a careful modern educationist, S. de Giacinto, views the problems we have treated here in general, in the following terms. "The true personality which is integrated in the intellectual function is a dynamic and not a static personality. And we must conclude that concepts, situations and intellectual integrations do not repeat themselves: they can be used operationally only once, and then the system must be reconstructed. In this way we can also understand that the dynamic of civilization consists in using up what has already been thought. The quicker the change, the shorter the models last, and this requires the preparation of new equipment." Further on, he adds: "So that the system shall not break up into thousands of fragments ... it must always remain a system ... in which the rules are constant. And since we have seen that nothing can be repeated, it is only constant obedience to syntax — an obedience which constitutes one of the two aspects of the creative capacity — that allows the intellectual system to remain a system in which every individual can recognise common ground with another, as well as feeling himself different." (From Chapters VIII and IX of the work *L'educazione intellettuale universitaria*, La Scuola, Brescia, 1962.)

These extracts are of interest as showing how vast the domain of conventional science has become, and as indicating especially that conventionalism itself is not a peculiarity of the sciences, but invests the world of thought and the personality of thinking man.

axioms are both "beginnings", and do not differ other than by the natural precedence that the latter have over the former.

The general formulation of a theorem is as follows: given the conditions already shown, A_i, B_i, C_i, . . ., prove that S_i follows from them. It is clear that if A_i, B_i, C_i, . . . did not follow from preceding theorems, S_i would not be demonstrable. But since the premises of S_i derive from one or more demonstrations S_{i-1}, and these in turn from S_{i-2}, and so back to an initial enunciation which derives directly from axioms A, B, C, . . ., it is clear that the whole chain of successive deductions would be valid if, instead of the original series of true axioms, A_1, B_1, C_1, . . . (which derive from A, B, C) had been posited axiomatically, or if the starting-point had been A_2, B_2, C_2, . . ., and so forth. The fact is, therefore, that every theorem can be considered as the first of a series of deductive developments, provided that the premises on which it is based are taken as axiomatic — although they themselves are derived, in general, from true deductive chains.

Given, in fact, that for the original axioms A, B, C, . . . neither certainty nor evidence is required, this is also true for A_i, B_i, C_i, . . ., constituting the premises to demonstrate S_i; hence this confirms that the deductive chain — naturally in its schematic form — could theoretically begin from any theorem whatsoever, positing axiomatically the conclusions of the theorem or theorems to which it refers.

522   Extensions to the factual disciplines

This point of view is important not for mathematics and logic, but to create a perfect parallelism between the syntax of the abstract and the factual disciplines.

It is justifiable here to remind the reader that many of the subjects we will discuss in this chapter have already been touched upon in preceding pages. Although perhaps getting an impression of tiresome repetition, the reader should evaluate these arguments for what they are, that is, as intended to show how the ternary scientific process of hypothesis–deduction–induction fits into the formal unifying structure of statistical method.

We recall that the theories of the factual disciplines are often derived from remote initial premises — whether theoretical, conventional, psychological — or from some other source. Braithwaite frankly doubts whether we may not have lost sight of the original postulates of many scientific doctrines, and there is nothing improbable in this. In the intellectual climate of today, who would care to trace back how the word atom (from *atomos* = indivisible) shed its primary meaning, seeing that the atom is not indivisible? Other theories have too recent an origin for their axiomatic and factual premises to have been forgotten — unknown as they were even at the beginning of the present century. But in the rapid evolution of thought, some of these premises have already found their way into the museum of historical relics. For example, we recall the case already considered (section 251) of economics, in which, after the Keynesian revolution, the classical axioms were edged away into the shade, giving place to the new principles from which macro-economics was developed.

In the majority of cases, however, the new theories start from old ideas; they do not constitute the first developing link of the original axioms, although they are based on starting-points already demonstrated or accepted, even if remote. In short, they correspond to the syntactic form of a theorem, the demonstration of which does not formally impose the certainty of the truth of the premises, or at least of their antecedents.

Returning to section 3221, we may put the hypothesis stated there in the form of a theorem. "Given two atoms of H and one atom of O, show that one molecule of water can be formed from these, without residue." It has already been shown that the demonstration is in practice impossible. But if somebody undertook to research into it, he would have to obtain the necessary information about the basic ideas of chemistry (elements, compounds, atoms, molecules, Avogadro's hypothesis, etc.) which are necessary to attain the result, without being in the least concerned whether they are axioms — and hence original assertions undemonstrable and perhaps not evident   or conclusions from antecedent theorems. Analogous considerations could be repeated with regard to the hypothesis implied in section 3225: "show that the oscillations of the pendulum are isochronous", but we will not dwell upon this at the moment.

523   True, false, and probable in the factual disciplines

Evidently, as with every mathematical theorem, a scientific theorem can be true or false, and the question hingeing on it can receive the answer yes or no.

But in science there is also a third solution, and that is the solution in terms of probability. This does not contradict the principle of the excluded middle, because factual questions, although sometimes having this form, do not always fit into the three perfect schemes of Aristotle[1].

---

[1] Cf. H. Scholz, *Storia della logica*, Silva, Milan, 1962, p. 30.

Besides, neither the yes nor the no of the sciences is apodictic, because phenomena are confirmed or denied by means of classifications established for them within the limits of accepted variation and on the condition *ceteris paribus*; that is to say, we neglect indefinite circumstances which in the affirmative case are not repeated, or we admit those which, in the negative case, happen nevertheless. One could quote a well-known Italian proverb: that against the force of facts reason doesn't count.

53   Syntactic analysis of statistical induction

For the reasons developed in section 37, the ternary process hypothesis–deduction–induction is by nature cyclical, and therefore leaves the choice of phase free when commencing the syntactic analysis for a thorough investigation of the formal structures of Statistics. It is convenient here to start with induction because a few remarks will suffice, this being the theme on which we have insisted in Chapters 3 and 4[1].

The concrete development of inductive work is certainly more complex than one would expect from considering it in the abstract: its roots go down to the origin of man himself (this refers to the biblical tradition, section 32), and even considerable ethnological and psychological inquiries would not suffice to lay them bare.

Thus, as Piaget[2] writes, Russell maintains that to perceive a white rose is equivalent to envisaging simultaneously the ideas of rose and of whiteness, that is, of an object and of a universal (identified in these pages with a "sensation" of Mach, cf. sections 171–172). One must recognize in this an arbitrary position, because "rose", although a very ancient and primary idea and word, is a common noun, and no different from the latest term invented for a novelty such as antiproton, penicillin, thalidomide, etc.

531   The formation of "cases" and of statistical data

Thus we come to the idea of classification and to the unifying process carried out on elementary observations to rid them of their individuality, that is to say of their historical character. In so doing we reduce them to "statistical cases", that is, to elements identical with others that can be grouped and enumerated with them to form those wholes which have already been called "statistical data" (sections 46

---

[1] The line of development of this and the following sections is taken from Chapters 3 and 4 of M. Boldrini, *Statistica, teoria e metodi*, Giuffrè, Milan, 1962.   [2] J. Piaget, *op. cit.*, p. 29.

and 462). A statistical "case" is therefore a unit occurrence of an event, and the statistical fact of the event makes clear the number of the units which compose it or which have been observed (sections 462 and 473 *et seq.*). If we say "Aldebaran" or "Betelgeuse" we are speaking of historical entities, and if we say "a star of the first magnitude" we are speaking of an event which becomes a statistical fact in the proposition "20 stars of the first magnitude".

Textbooks generally pass over in silence the formation of cases; and philosophers do likewise when they come to admit that "rose" is an immediate datum, almost a natural fact. The truth is that without starting from the formation of cases, there can be no induction: here begins the creation of the uniformity of nature by the human mind, from which are produced the structures of every factual regularity.

For science, then, there are neither preconstituted cases nor classes: every original inquiry returns to the point of origin, every considered induction must needs revise and work out again the cases on which it proposes to operate. It may happen that in breaking down and building up again, one arrives at a confirmation of the classification already existing. In spite of this, the result is always an achievement because it adds to the content of previous knowledge and can promise fresh discovery. The evolution of the meaning of the noun "water" (on which we dwelt at some length in sections 3553 *et seq.*) is a particularly cogent confirmation of this statement.

Earlier it was said that concepts are more numerous than things (section 46). Nothing is more true, because the same individuals can be classified according to different criteria in an unlimited number of categories. Thus, if someone said "white rose", one could ask if he was speaking of a botanical fact, or an ornamental object, or the reverence of a little girl at the altar, or of the plant whence comes attar of roses, or of a feminine name. It was also said that, according to the point of view, all botanical entities with a flat-topped inflorescence (section 381), such as marguerites, alpine stars, thistles and sunflowers, can either form a single large family, the Compositae, or can be broken up into innumerable transitory "species".

Hence it is not surprising that the requirements of science lead to the reclassifying of "events" which practical needs and earlier research had classified differently.

5311   The provisional nature of classification: new botanical species

To enable us to consider closely the variety, wealth and fruitfulness of the human classification of natural events, so different from the actuality of Mach's "sensations", we will consider some examples, starting with a very simple case taken from the gardening column of an English magazine.

"Mahonias were at one time included in the family Berberidaceae and in some catalogues are still shown as such (they are so listed in the *Italian Encyclopedic Dictionary*). Now, however, they form a genus apart, and they are distinguished from the barberry genus by many characteristics: they are invariably evergreens, with few or no thorns, and the leaves terminate in a single prickle. . . . About forty species are known . . . and the commonest is *Mahonia Aquifolium*. It is a shrub originally from north-west America, and in English gardens grows to a height of 3–4 ft."

Thus, "Mahoniaceae" is a new term invented by gardeners, in every way analogous to the ancient word "rose". Mahonias, says the description, are distinguished by "many characteristics", that is, by characters at first completely overlooked and which, *ceteris paribus*, caused them at one time to be considered identical with Berberis.

Here is another example. In 1949 Professor F. Lona, during his excursions in the Orobian Alps near Lake Como, identified a new botanical species growing on the shingly slopes of the Pizzo Arena at 2,000–2,400 metres. It was originally confused with *Linaria alpina*, or, on less authority, with *Linaria thymifolia*. Lona accurately described the species in the traditional Latin of the naturalists' system, differentiating it from kindred species, and after stating "Clarissimo Botanico Sergio Tonzig hanc speciem dedico", he gave it the name of *Linaria Tonzigii*.

5312 The provisional nature of classification: syntactic rules

Language is one of the most complex of human phenomena and it would need a philologist to extract from it illustrations suitable for the subject under discussion here. But the elementary ideas shared by everyone are enough to remind us that the grammar of a language is nothing more than the characterizing classification of its expressive moods[1].

Thus, the rules of syntax learnt at school teach — let us say — that in Latin, when the nominal predicate is an adjective (*bonus*), it agrees with the subject in gender, number and case; when it is a substantive

---

[1] On the classificatory nature of definitions and grammatical precepts, see F. W. Householder, "Lists in Grammars", in *Logic, Methodology and Philosophy of Science*, Stanford University Press, Stanford, 1960, especially p. 571.

(*faber*) it always agrees in the case, except for nouns which have a masculine or feminine form (*magister*, *magistra*), which, however, agree in gender and number. "Rule and exception", said the grammarians. But the syntax rule is straightway generalized when it is extended to affirm that "the nominal predicate always agrees in gender, number and case". Obviously, it should then be clear that nouns not varying in number (*divitiae*) or in gender (*domicilium*) do not follow this rule.

These remarks bring us back to the generalization that every change of classification, even in grammar, determines variations of the contents of the classes and shows up their purely subjective nature. In the botanical "events" considered in the preceding section we were dealing with limiting classifying variations and hence with transitions from genus to species, while the grammatical examples examined above are concerned with broadening the classes, thus constituting transitions from species to genus (section 322).

5313 The provisional nature of classification: strange particles

More complicated situations arise, not so much when (as in the preceding sections) the evidence suggests the need for new classificatory compartments, as when changing circumstances require revision of the model hitherto accepted. This condition has already occurred in section 472, in connection with the Lexian distribution of the proportion of the sexes in human births. Here are two important new instances.

The first of these concerns the discovery in the 1950's of certain elementary particles, divisible into two groups, namely K-particles ("heavy mesons") with a mass less than that of protons, and "hyperons" with a mass greater than that of protons. In order to distinguish these from other particles, a new quantum number was introduced, called "strangeness", differing from zero only for the strange particles. One of the major theoretical difficulties about these particles is that the duration of their half-life (which has limits from 10^{-11} to 8 seconds) is longer than would be expected from theoretical calculations based on the frequency of their production. It is also difficult to interpret the fact that in the interactions between particles the strangeness is maintained in some circumstances but in others not; and that in particular cases (interactions between nucleons or between π-mesons and neutrinos) there is an "associated production"; that is, one single strange particle can never be produced, but two must be, and the sum of their strangenesses is equal to zero.

The spin of heavy mesons is equal to zero, but not that of hyperons; as we have said, the half-life of both varies, just as their decay-times and other characteristics do.

Evidently, the question of where to place the strange particles has disturbed pre-existing systems, and their position in the scheme just mentioned depends on further discoveries.

Extending this concept of "strangeness" to other facets of the physical world, we find ourselves talking nowadays of an *étrangeté de l'univers* which, after all, expresses nothing if not the perennial difficulty of containing the conflicting facts of the world in a rational system[1].

5314   The provisional nature of classification: inflationary economic recession

In the economic field, the doctrine has long been affirmed of the connection between inflation and prosperity and between deflation and recession.

During the last 150 years, in the United States, there was inflation in 17 of the 20 main periods of prosperity and deflation in 18 of the 20 most obvious periods of recession. In no case, however, even in the exceptions, had there been inflation in a recession ("inflated recession", according to the name adopted by Hans Apel[2]). But in the recession of 1957–8 the paradox did indeed happen: in the United States, between July 1957 and April–May 1958, the gross national revenue, industrial production and employment underwent a considerable decline, while the index of wholesale prices increased by 0·9 per cent, the index of commodity prices increased by 2·2 per cent, and wage rates per hour in the manufacturing industries also increased by 1·9 per cent.

These facts, contradicting a rule tested over a long period, led to a revision of the whole theory of cycles and to the question being posed

---

[1] These ideas derive from an article by E. Fiorini, "Particelle elementari", in *La Scuola in Azione*, 1960–61, No. 16, while in the study by G. Lemaître, "L'étrangeté de l'univers", contained in the small volume. "Un nouveau système de chiffres et autres essais", in *Quaderni della Scuola Enrico Mattei di Studi Superiori sugli Idrocarburi* of ENI, No. 10, 1961, the "strange particles" are also mentioned, but only to show that the whole of modern physics is rich in étrangetés, so that they do not constitute a peculiarity of the more recently discovered particles.

[2] H. Apel, *Inflated Recession, a New Economic Paradox*, The C. K. Kazanjan Economic Foundation, 1958.

whether inflation was not becoming a constant element in the economic system, and whether its presence had not till then partly escaped the vigilance of economists. There followed an intensified study of the price mechanism, and revised views about the theory.

Thus, among the principal symptoms of the phases of expansion and recession it is no longer acceptable — as was held until the last recession — to include inflation and deflation. Other frameworks are needed which agree with the emerging facts.

5315   The provisional nature of classification: evolutionism and Mendelism

A similar lesson can be learnt from a discovery made over a century ago (section 463) and whose implications, originally unforeseen and unforeseeable, are gradually coming to light. In 1865 the Abbé Gregor Mendel, already a biologist of repute, presented to the Society of Naturalists of Brunn a paper with the modest title: *Versuche über Pflanzenhybriden*, but practically no-one paid attention to it. In 1859 Darwin's *chef d'œuvre*, *The Origin of Species*, had been published: the scientific world was in a turmoil, and the interest of naturalists was being concentrated on the differences between living species and on small individual divergences. A particular consequence, well known to statisticians, was the investigation into the inheritance of biological characteristics by Sir Francis Galton, a talented cousin of Darwin, who, in a work published in 1889, showed that, in human groups, height is inherited from the whole chain of ancestors according to rules whereby the children of tall parents as well as those of short parents tend towards an average height.

A quarter of a century earlier, however, Mendel, unknown to Galton, had solved the problem in a radically different way, and so conclusively that the Englishman's statistical exposition is now regarded as of little more than historic interest. Instead of investigating small variations — a method favoured by the Darwinians — Mendel had crossed plants differing completely in one characteristic (such as peas with smooth or rough seeds, with green or yellow cotyledons, with long or short stems, and so on) and had discovered inherited factors which no-one had previously noticed. All that had happened was a change of criteria, aided by a touch of genius.

When, at the beginning of the present century, the memoirs of Mendel were re-discovered, an impetus was given to the science of genetics. It was shown that species are indeed subject to statistical variation, but that accidental variations are not inherited; that, of the opposite characteristics of parents, only one is transmitted to the first hybrid generation, while the other reappears only in the second generation in 25 per cent of the cases (section 463, 471 and 4711); that species do not vary gradually but change suddenly; and that natural selection working on the mutations contributes to the formation of new species. This discussion shows that species can be classified by modifying the defining characteristics in an unlimited number of ways.

Taking the present case with those of preceding sections, we must agree that in every field of the natural sciences — and not only in biology — variation of the classifications, determined by new discoveries or by changed needs, can revolutionize not only the grouping of a species (section 5311) but also the most complex and accepted doctrines. We shall shortly meet with the confirmation of a less momentous discovery, connected with a world-famous name (section 5431).

54　From induction to hypothesis: the intuition of genius

After induction (section 53) we now pass on to consider hypothesis, of which we have already had a glimpse (section 313). Hypothesis in general is not easy to discuss, because the tentative formulation of a scientific generalization is often arrived at by a stroke of intuition which, by its very nature, will not reduce to a system. Galileo thought thus, as Harrod and others have pointed out. The well-known description by Vincenzo Viviani of Galileo's discovery of the pendulum is relevant here. "Around the year 1583, when he was about twenty, Galileo found himself in the city of Pisa . . . and one day in the cathedral of that city, curious and perceptive as he was, it came into his mind to watch the motion of a hanging lamp. . . ."

Again, Colerus tells us that because of his renowned mathematical gifts, Leibniz had been asked by friends of Pascal to sort out the documents left by the latter, and in carrying out this task he came across a graph of the sine function in the first quadrant. In the receptive mind of Leibniz this drawing was generalized, until he recognized in it the so-called "characteristic triangle", from which is derived differential calculus.

An instructive and singularly prophetic remark by a seventeenth-century German naturalist, Athanasius Kircher, was quoted by Röntgen in the rectoral discourse of the University of Würzburg in 1894, reprinted some years ago. "Nature at times presents stupefying marvels in the simplest things, but these come to be known only by persons who, with insight and a keen intellect directed towards research, allow themselves to be advised by the mistress of all things."

A modern commentator on this passage points out that, in remembering it, Röntgen seemed almost to presage his own great discovery of X-rays, then imminent. And years before Röntgen, Pasteur had written: "Notice that in the field of observation, what is called chance only favours alert minds."[1]

541 Scientific explanation

To solve a scientific problem posed in a hypothesis always means to "explain" something, to reveal to the mind an area of the world still hidden; in a word, to unfold and to read the scroll which contains the secret of inference S.

In ordinary language the word "discovery" is used to indicate the making known of a natural fact or an experimental result of great importance. But in the field of science a discovery is anything which throws new light on the ternary process of research and helps to enrich the patrimony of knowledge. To reveal the hidden S by demonstration of a proposed theorem is therefore always to make a discovery.

5411 Explanation is a formulation of the *quia*

But what is this secret? It should be clear to anyone that science today no longer seeks those "essences" which Galileo had already rejected (section 39), nor those objective and definitive "truths" and those "causes" which led positivist metaphysics astray. In today's climate of scientific relativism, which makes any description precarious and attributes a historic dimension to every factual interpretation, we could well turn to the wise and almost paternal advice of Dante (*Purgatorio*, 3, 37–9) (reading it, of course, with a modern eye):

> Rest content, race of men, with the *quia*;
> For if you had been able to see all,
> There was no need for Mary to give birth;

where *quia*, as Sapegno points out, is used in the sense that it had in scholastic language.* Benvenuto Cellini makes a similar comment:

---

[1] The extracts from Viviani and Pasteur are quoted by M. Boldrini, *Statistica, teoria e metodi*, pp. 83 and 93, note; for that from Kircher, see *Zur Geschichte der Physik an der Universität Würzburg*, Festrede von Dr. W. C. Röntgen, z. Zeit Rektor, Würzburg, 1894, reprinted 1959.

* *Translator's Note* — These passages are difficult to render into English because of the distinction in mediaeval philosophy between *quia* and *propter*. Broadly speaking, *quia* concerns the relations between observed events, *propter* their underlying causation.

"Sufficiat vobis credere quia sic est, et non quaerere propter quid est, in his de quibus non potest plene assignari ratio vel causa," etc.

With equal wisdom St Thomas Aquinas had previously written: "Rerum sensibilium plurimas proprietates ignoramus; earumque proprietatum, quas sensu apprehendimus, rationem in pluribus invenire non possumus" (*Summa contra Gentiles*, I, 3)[1].

Science, therefore, must try to observe things as they are and not their reasons — the functional relations and not the causes which could not be established *plene*, without the Christian revelation.

5412 Pragmatic truth and *noscendi cupiditas*

But, it is said, not only have all the absolutes, the values, and the finalisms been banished from modern science but also the objectivity, in order to make way for pragmatic truth — that is, for logical consistency and functional opportunism. Therefore, to know the *quia* cannot mean to describe in a passive manner an objectivity already rejected *a priori*, but rather to re-combine and integrate every piece of evidence in the widest possible range of ideas. It is perhaps this comprehensiveness which corresponds to the popular idea of a "great discovery" (section 3561).

Cournot observed that the transcendence of π disappears in the ratio of the surface area of a sphere to its greatest circle, because $4\pi r^2/2\pi r = 2r$ is a perfectly clear result. Contrariwise, introducing a complex number into the formula for solving the quadratic equation $x^2 + 2bx + c = 0$, far from complicating things, unifies the concepts in a practical way and in a sense clarifies them. In fact, if we put $x = -b \pm \sqrt{(b^2 - c)}$, solutions (real or imaginary) can be obtained whether the difference under the square root sign is positive or negative.

Now, the major part of scientific explanation consists in finding lines of argument which, if not of this kind, are always such as to

---

[1] And again: "In Astrologia ponitur ratio excentricorum et epicyclorum ex hoc quod hac positione facta, possunt salvari apparentia sensibilia circa motus caelestes; non tamen ratio haec est sufficienter probans, quia etiam forte alia positione facta salvari possent" (*Summa Theologica*, I, q. 32, a. 1 to 2). And elsewhere: "Illorum [astronomers'] suppositiones [Ptolemaic system] non est necessarium esse veras. Licet enim talibus suppositionibus factis apparentia salvarentur, non tamen oportet dicere has suppositiones esse veras, quia forte secundum alium modum, nondum ab hominibus comprehensum, apparentia circa stellas salvantur" (*De coelo*, L. II, lect. 17). These and other quotations from Aquinas are given in M. Boldrini, "Causa vel casus", in *Statistica*, **20**, No. 4, 1960.

satisfy the *noscendi cupiditas* — the torment, and at the same time the benefit, arising in the mind of man from the Renaissance onward. With this and because of this, the solution of the problems contained in hypotheses brings its own reward, though it is at once nullified by persistent new doubts, giving rise in their turn to new truths.

542   Hypothesis and the enunciation of a theorem

Re-examining the general idea of a hypothesis (sections 313 and 54), we see that — as with mathematical theorems — it can always be reduced to the task of researching whether and in what way the proposition S is implicit in the propositional premises A, B, C, . . ., and can become explicit. In other words, it is a matter of referring the answer back to the question contained in the hypothesis, and — in the absence of experimental proof — of demonstrating its rational and factual consistence with the premises.

From the relation between axiom and theorem (section 521) there emerges an analogy between hypothesis and theorem. In fact, a hypothesis is a statement to be demonstrated and an axiom is a statement to be accepted; hence at least their common position of "beginning", in the sense set out above (section 52).

5421   Answering a hypothesis by means of analogy

As we have seen, the scientific demonstration of an assumption consists in referring an unknown phenomenon to the sphere of action of one or others already known. Someone who says that a flash of lightning and a thunderclap are the manifestation of an electric discharge in the atmosphere has explained the phenomenon. But, again, one who draws an analogy — for example, St Paul in saying that the baptized are members of the body of Christ (I Cor., 12) — also gives an explanation, in so far as people are satisfied with it.

Having established this, let us consider again the isochronism of the pendulum. Galileo discovered it when young and profoundly interested in the laws of motion; realizing that a pendulum oscillates for the same good reason that a weight falls, he had found a first solution to the problem. This step led on to others, and only with Huyghens was found the famous formula with two parameters,

$$T = 2\pi \sqrt{\frac{h}{g}}.$$

This finalized Galileo's investigations.

Towards the end of the same century, the seventeenth, Newton

H*

showed that the fall of weights is an expression of a gravitational force f, proportional to mm'/r^2, involving the masses of two bodies and the distance between them, thus placing the behaviour of the pendulum and many other phenomena, apparently contrasting, in a more general light.

Summarizing in rather haphazard fashion various pieces of knowledge connected with Newton's law, we can now say that not only falling weights and the oscillating pendulum, but also the motions of the planets, the rising of a flame, the vibration of a string, air bubbling in water, the rising of the mercury column in a Torricellian vacuum, the double cone which appears to run uphill on an inclined plane with a V-shaped notch, the rising and falling of the tides, the orbiting of electrons round the atomic nucleus, the emission and absorption of light, heat and of electromagnetic energy are all manifestations of Newton's principle. Clearly, each of these effects is the result of the demonstration of a scientific theorem, i.e. it is a discovery in the sense used above.

Considering them carefully, one sees that these effects are more than mere descriptions of things observed, but do not go beyond the limits assigned above to discovery and scientific explanation.

543   Answering a hypothesis by a more extensive description: cyclicality of marriages

The answer to the problem posed by a hypothesis can be found in a more thorough description of the phenomenon, as can be seen if we consider the variation in the number of marriages during the week.

It may be that a cyclical function like that of Fourier can be fitted satisfactorily to the empirical data, and that the resulting parameters constitute an adequate description of the daily variation of the phenomenon. In such a case, the description would be perfectly adequate. But, again, suppose that not only the original data but also the associated mathematics revealed that the fewest marriages occurred on Fridays and Tuesdays and the most on Saturdays (Sundays in Italy). If we link this with the known popular preference, the description of the phenomenon will be more complete, without having in the least gone beyond the *quia* to the *quod*.

The reduction of the oscillating pendulum and the other phenomena mentioned above to the notion of a falling weight is similar in principle.

5431   Answering a hypothesis by an axiomatic solution: a discovery of Pasteur's

An autobiographical passage of Pasteur's (see, in his *Works*, the reports of two conferences held at the Chemical Society of Paris on the

20th January and the 3rd February, 1860) shows that doubt concerning a hypothesis can sometimes be eliminated by a statement which is indisputable only because it is accepted axiomatically. Pasteur relates the discovery (made at the age of 26!) of the relation between the rotatory power and crystalline forms of the tartrates and the paratartrates of both sodium and ammonia. Since 1848 it had been a puzzle that, of two substances physically and chemically identical, the tartrates turned the plane of polarization towards the right, while the paratartrates had no effect on polarized light. Pasteur began his researches almost by chance, and ended by asking himself whether the solution to the mystery did not lie in a different type of crystallization. This was the hypothesis.

Pasteur then discovered that all the crystals of the tartrate were of one kind, while those of the paratartrate were of two types, symmetrical in form, like a pair of hands. When they were separated and two solutions prepared, Pasteur was able to state, "with surprise no less than pleasure", that the first solution caused polarized light to deviate to the right and the second to the left, while a mixture of equal parts lacked all power to rotate. From this celebrated experiment two things emerged clearly:

(1) That is was enough to shift research from the plane of chemical analysis to that of crystalline forms, to get to the root of an apparently insoluble puzzle. (Similarly, by paying attention to contrasting characteristics which had previously been neglected, Mendel discovered the first laws of heredity (section 5315).)

(2) The satisfying explanation which Pasteur gladly accepted did not depart from the *quia* rule. Polarization occurs in two ways because the crystals concerned are of two types. As soon as one obscure phenomenon had been linked up with another, Pasteur was little inclined to ask himself *why* the solution of hemihedral crystals polarized light in opposite directions. He accepted this fact, no less mysterious than the other, as an axiom and used it to demonstrate his theory.

55   Deduction in the sciences

It was not easy to separate the previous explanation of the nature of a hypothesis (section 54 *et seq.*) from the discussion which now follows concerning the deductive phase, in which we inquire into the syntax of deriving a theorem from the proposed hypothesis. This is because it is difficult to examine the structure of an inquiry without becoming involved in the possible solution. As in a chess problem, the inquiry follows certain rules, and formulation of the problem helps to solve it. In science the point may be emphasized because the propounder of a theorem will often be demonstrating it and, consequently, far from setting one side against another — as in the combat between Œdipus and the Sphinx or in the imaginary chess-game — he alone is involved.

551   The result of deduction demands experimental control

The deductive phase initiated by the question posed with a hypothesis searches for the truth that is called "scientific law". The question, "perhaps the pendulum is isochronous?" which Galileo asked himself in 1583 in the cathedral of Pisa, is a hypothesis; the answer, which could have been given by him and others after long mental and experimental labour, "yes, the small oscillations of a pendulum are approximately isochronous", is a law; and the formula in section 5421 is the "model" of it.

One can say, then, that the transition from hypothesis to law and model consists in formulating the system of propositions which in a science come to be recognized as true after a certain time (as Carnap puts it — see section 17 — though in a very broad sense); it consists, that is, of a deductive development whose conclusions would have to be checked by experiment (section 3221 et seq.). Since the hypothesis is inductive in origin and since the next deduction requires further experimental and generalizing checking, it can be seen that the entire procedure is of a repetitive nature (section 37) and does not stop until the problem is solved, that is to say — according to the discussion in section 5421 — until a satisfactory explanation, or law, is found for it.

5511   Law and theory

The word "law", repeatedly used above in the sense of "rule", is the goal of every piece of scientific research based on the ternary process hypothesis–deduction–induction (sections 313, 3221). Some people, in theorizing, hold that a definite distinction exists between law and theory, and they reserve the first term to indicate directly tested regularity, using the second to designate those regularities which are derived only indirectly from observable phenomena. The word "hypothesis", it is said, can be used to mean a law not yet sufficiently documented, but above all "it acquires its full significance with the theory, because here one actually assumes something never yet perceived".[1]

---

[1] A. Maros dell'Oro, *La teoria fisica*, Cedam, Padua, 1955, p. 10. It is not that this author insists on the distinction so drastically, but that he constantly uses it, at any rate as an explanatory guide.

This is a respectable and tenable point of view, but not one which could be adopted when (as in these pages) one admits subjectivity, which simply requires the consistence and plausibility of every scientific formulation. It must, of course, be recognized that some laws, at least at first sight, seem clearly demonstrable (e.g. the fixed boiling-point, which appears to be verifiable but in reality is proposed as a limit); while many theories have an essentially hypothetical form (for instance, the origin of the solar system, which can be described as in the Book of Genesis, or according to the theory of Kant–Laplace, or that of Jeans or of Jeffreys, and others).

But one senses at once that the presence of interpretative elements in every scientific formulation reduces the difference between law and theory to a simple question of degree. It is sometimes held that even statistical information is always indirect[1] because one does not take into account historical elements but only symbols, such as numbers, tables, and questionnaires[2].

Here one may add that even the most self-evident of laws is the indirect expression of facts about the world stated scientifically.

As was said above, rather than distinguish between hypothesis–theory and law, I prefer here to keep hypothesis separate from law (or from theory if the term is preferred). And I insist on giving the name "hypothesis" to every scientific intuition awaiting documentation and confirmation (section 313). It is true that the formulation of a hypothesis also implies some degree of observation, and in this sense dell'Oro's comment is to be accepted; but for the hypothesis to become a law, experimental or rational confirmation is needed to satisfy both the scientist's desire for knowledge and the demands of his critics.

From this a formal corollary can be drawn that today a scientific result no longer takes on an objective meaning, to be understood as a reality that is external to man and his ideas.

Of necessity, the relevant law (or theory) should no longer be expressed in an apodictic form (for example: given the circumstances a, b and c, it is asserted that K is the cause of the phenomenon observed) but only as a truth consistent with the actual state of knowledge (for example: given the circumstances a, b and c, the observed phenomenon is interpreted as functionally connected with the system K). In science,

---

[1] For statistical information regarded as "always indirect", see M. Boldrini, *Statistica, teoria e metodi*, p. 120.

[2] On the significance of questionnaires see M. Boldrini, *ibid.*, pp. 111 and 221.

form is important; but the fact that scientific results may be expressed in the first of these two modes (as often happens) will be a minor defect in the eyes of a reader who knows that only the second expression is rigorous.

5512   Cautions against the assumption of general principles

In this connection one must take care not to reintroduce outdated apriorisms. Going back to Bergson and Meyerson (and he might also have mentioned Mach as a forerunner), dell'Oro reminds us that certain statements are equivocal[1], such as Ampère's "the mathematical laws of motion of the stars have regulated such motion since the world existed, and long before Kepler demonstrated them". Although the myth so phrased by Ampère has now been discarded, thousands of scientists still quietly believe in it, as dell'Oro laments. But an exact formulation consists not in the correction suggested by him — "the motions of the planets were regularly elliptical long before Kepler, and the latter only pointed out and summarized such regularity in such laws" — but in this other: "before Kepler there were in science no elliptical movements of the planets but only the circular motions of Copernicus; indeed, for those who did not believe that the former were yet demonstrated, the circular motions of the sun and planets, with their epicycles, round the earth were true. According to the later formulation of Kepler, the planets now describe (as science understands them) elliptical orbits with the sun as a common focus." Only with this pronouncement — which is also in harmony with the thought of Pearson[2] — can it be made clear that Kepler's laws are not simply a

---

[1] A. Maros dell'Oro, *La teoria fisica*, p. 84.

[2] K. Pearson, *The Grammar of Science*, Dent, London, 1937, p. 36. The formulation proposed in the text no longer seems unacceptable when compared with another case that runs parallel but is more consistent with commonsense. We refer to the argument of Galileo with Father Grassi (section 3224). Scientifically, the facts discussed should be set out as follows. Before men achieved high velocities, it was scientifically false that eggs could be cooked by moving them rapidly through the air; on the contrary, they only cooled down. But with high experimental velocities, eggs (scientifically speaking) can be cooked by the attrition of the air. Here, what is true or false for science is likewise so for ordinary experience, and hence the formulation engenders no incredulity. But the agreement is purely fictitious, because in science one pays attention to the rational interpretation of phenomena, while when reasoning from common sense one takes account only of physical possibilities. Therefore, in the sixteenth century, Galileo was perfectly right and the claim of Father Grassi, far from being enlightened anticipation, was simply in error.

description of facts, but a concept embodying a number of decisions (for a similar question see sections 3554 and 3555).

552 Jeffreys' principle of simplicity in Nature

In section 336 we mentioned the origin of the idea of the simplicity of Nature, and in the following section 34 we set out Jeffreys' theory of induction, without, however, going into detail because at that point we had not introduced the idea of probability. We will now complete our discussion of the subject, considering the position which the British astronomer takes up with regard to deductive inference and to what he calls the "postulate of simplicity". Jeffreys begins by referring to the theorems of the sum and product, Bayes' theorem (section 4321), and other aspects of probability, and he proposes to demonstrate two theorems which will not be fully developed here because somewhat extraneous to the discussion[1].

With the first of these, Jeffreys shows that, given a hypothesis q, which determines a series of observable consequences $p_1, p_2, \ldots, p_n$, when H is the initial knowledge of the phenomenon, the probability that the hypothesis is true gradually increases as the confirmations increase, because the probability of success of an nth test after $n-1$ favourable tests tends to unity. It is shown, in fact, that

$$P(p_n|p_1 \cdot p_2 \cdots p_{n-1} \cdot H) \to 1.$$

If the probability of q is zero, that is, $P(q|H) = 0$, however many tests are carried out it will not increase: but if it is not zero, it will increase with each new test (without necessarily becoming high) by reason of the preceding limit, in which — be it noted — the hypothesis q does not enter.

The second theorem says that if for a hypothesis or law q we have a measure θ which can be contained between a restricted interval ϵ when the hypothesis is true and between a much wider interval E when it is false — that is, when we have $\sim q$ — the relation between $P(q|\theta \cdot p)$ and $P(\sim q|\theta \cdot p)$ is influenced by that of $E|\epsilon$, and consequently "the more precise the inferences drawn from a law (q), the greater is the probability of its being verified, although the contradictory law ($\sim q$) also admits of a forecast compatible with observation"; that is, it is not entirely excluded. The first of the two theorems turns out to be linked in a certain sense with that of Bayes (to which Jeffreys constantly refers), because it shows that the probability of the given proposition q increases when repeated verifications confirm it; the second shows that

---

[1] H. Jeffreys, *Scientific Inference*, pp. 34 *et seq.*

if the limits between which the determinations θ of q vary are restricted, the hypothesis is more likely to be true than false.

Jeffreys draws from these two theorems a whole series of consequences of which the following is fundamental.

The number of scientific laws, before there is any information H whatsoever, is necessarily great, but the sum of the respective probabilities — these being mutually exclusive — cannot exceed 1. Consequently, they can be considered as constituting a random variable, to each value of which a probability corresponds. Now Jeffreys states that the probability of each law varies in inverse ratio to its complexity, and consequently the most plausible laws are the simplest. Expressed in terms of probability, this is the "simplicity postulate" which in its most general form was mentioned and commented upon in section 34 (before the concept of probability had been introduced). The criticism applied to it then is valid here. There is no reason to suppose that the form of the laws of concrete phenomena, which present themselves in great variety in the course of scientific inquiries, can be selected in advance by abstract argument, on the basis of the even more abstract principle that their probability would decrease because of the increasing complexity of formulation. Add that it may be questioned whether this principle is consistent with the postulated independence of probabilities, which might not be the case with regard to two laws which differ by a parameter.

553   Harrod's principle of simplicity in Nature

Consequently it is not to be wondered at that Jeffreys' principle of simplicity is rejected by Harrod who, while reaching the same conclusion — that is, of preferring the simplest expression of a law — uses arguments which, from his point of view, are much more pertinent[1].

Starting from the usual probabilistic syllogism: if P is generally (or is not generally) followed by Q, an observation of P is probably followed by Q (or is not followed), Harrod proposes first of all to invoke an inverse principle called "reversing the consequents" which he enunciates as follows: it has been observed in a set of cases that P is followed by Q, then the hypothesis that P is generally accompanied by Q is more probable than the contrary hypothesis. Without entering into Harrod's lengthy discussion and to come at once to the criticism directed at Jeffreys, Harrod argues:

---

[1] R. F. Harrod, *Foundations of Inductive Logic*, Macmillan, London, 1956, pp. 86, 149 and elsewhere.

(1) One must not assume that a simpler formulation of a law has superior prior probability, because there is no basis whatsoever, outside of experience, for maintaining that the universe is constructed on simple lines (section 336). "Simplicity may have an aesthetic appeal; simple laws may appeal to us on utilitarian grounds; but all this has nothing to do with logic."

(2) Before an inquiry or before empirical evidence, no law has any probability whatsoever; hence neither a simple law nor a complicated one.

Jeffreys, adds Harrod, affirms that simple laws allow of better forecasts (cf. section 34), and that this would be an excellent reason for preferring them. But how can one know this in general? Jeffreys proposes as an example that six experiments with a solid sliding down an inclined plane corresponded, within the agreed limits of error, with Galileo's law $x = \frac{1}{2}at^2$, where x is the distance in centimetres from the origin and t the time in seconds; and he recognizes that the same result could be formulated in an infinite number of other ways. But if the universe were completely irregular — or, as Harrod puts it, if it were Heraclitean (cf. section 347) — in very many experiments the same distance would be covered in many different lengths of time, equally frequent, and hence among them there would be some instances conforming to Galileo's law.

Consequently a series of six experiments, all, so to speak, Galilean, would not entirely exclude pure randomness. If the event has happened, however rare it might be in a Heraclitean universe, there is nothing to do but to take note of it and recognize that, with respect to other results, it reflects a simple law. Will it on this account be the most probable? Not on Harrod's hypothesis: he would derive the result from the principle of reversing the consequents.

But, Harrod continues, if the universe were perfectly Heraclitean, not only those six but every other assortment of six observations would be equally rare. If not, among all the rare groups, the six supposed observations present the peculiarity of corresponding to Galileo's law, while the others — where one can find a rule in them — would correspond to laws at least more complicated. Then Jeffreys' supposed experiment would lead one to think that the universe does not behave in a purely random way, and this conviction proceeds from the simplicity of the law which it expresses. This, and only this, says Harrod, can be the guiding criterion for preferring, in defining a phenomenon, a simple law (even if approximate) to any other more complicated one, although exact in the long run. This is because, in a chaotic universe, the approximate verification of a simple law is more improbable than the verification of a complicated exact law. With an ordinary pack of cards one can find as many examples of this point as one wishes.

56   Simplicity not a principle of Nature but a human assumption

No doubt there will be partial agreement with Harrod's objections to the views of Jeffreys: but Harrod, as already mentioned, admits that there are many uniformities in Nature and that the laws describe those uniformities. Now this premiss — be it explicit or implicit — could not be accepted by anyone who has followed the theory of induction developed in this book. If one admits that natural regularities exist, one is accepting an *a priori* principle not recognized by current epistemology — an attitude which is authoritatively formulated in the following two axioms of Wittgenstein:

6.363 The inductive process consists in assuming the simplest law that can accord with our experience.

6.3631 This process, however, has no logical justification, but only a psychological one.

It is, in fact, clear that there is no reason to think that it is the simplest case that happens.

Furthermore, we must admit that Nature is what it is, and only the Creator has penetrated its secrets. Nature as known to science is based on elementary observations, grouped in homogeneous classes on the basis of intrinsic characters, made independent of time, and embodied in the cultural field to which the theories refer. Of those theories, static in themselves, the scientific laws are part, and these do not emanate from the facts but are imposed on them for their understanding and utilization. This concept, still absent from the positions of Jeffreys and Harrod (despite the anticipatory statements of Mach on the economic basis of scientific concepts, section 311), deserves close attention in order to complete the idea of Statistics, which has gradually been asserting itself and was summarized in a more formal manner at the end of the preceding chapter.

561   No general principle is necessary to science

To seek a purely rational co-ordination of a complex of events implies a tacit faith in some principle or concept analogous to those to which the earlier natural philosophers attached the first link in their deductive chains, as though to a hook: for example, not only evidence, or universal causation, or the regularity or simplicity of Nature, but also finalism, vitalism, the *vis medicatrix naturae*, sense-impressions, substantial forms, physical faculties, phlogiston, and then ether and gravity — to mention the two latest, illustrious victims of modern relativity theory. Among this rich and ever-growing collection of false ideas will also be found Jeffreys' assumptions such as "the probability existing before having any information whatsoever", "the general principle H of a theory", and "the postulate of simplicity", as well as Harrod's aspiration, not put into so many words, to find regularity in Nature perceivable through induction. Anyone who has seriously studied this book will be convinced that a world about which nothing is known originally must be built up by the human mind bit by bit, day by day, as ideas are gathered: and hence that it is useless to aspire to objectivity if understood as something external to man, or to probability if understood as something existing *a priori* and hence to be read in things, as once men hoped to read universal mathematics in all creation.

562   The parameters of scientific laws: their pragmatic basis

Speaking in general of scientific laws, Jeffreys states that many of them contain numerical parameters which are determined by counting, as when one says that "the mustard flower has six stamens", or by measurement, as is the case with the gravitational force f of Newton, or with the parameter g which occurs in the formula for the pendulum already mentioned (section 5421). These parameters, too, are clearly derived from the concept of probability.

To understand this we will take an example analogous to that of the mustard flower, namely a peculiarity of the medusa or jelly-fish, called *Pseudoclytia pentata* because of the number of radial canals commonly existing in its mantle. Supposing the species, its mantle and its radial canals (anything but simple and indisputable ideas) to be already defined, we propose to count the number of these canals. In a sample of 996 medusae, fished up "by chance" as in a lottery, the distribution of the number of canals is given in the following table:

Number of radial canals	Number of specimens	Proportional distribution
2	1	0·0010
3	8	0·0081
4	56	0·0562
5	860	0·8634
6	64	0·0643
7	6	0·0060
8	1	0·0010
Totals:	996	1·0000

From these numbers one draws the immediate conclusion that the adjective *pentata* applies to one part only of the medusae. Actually, from the above table one can derive the following points:

(1) The number of radial canals varies from 2 to 8. Statisticians call this interval the "range of variation".

(2) The propositional reduction of the frequencies to a total of 1, shown in the last column, corresponds to the "probability" of drawing a sample by chance with a predetermined number of radial canals. The numbers in the first column, read together with those of the third, form a "random variable" in which five canals have the greatest probability.

(3) The most frequent number of canals is, in fact, 5. This most frequent number, called by statisticians the "mode" (the most usual value), is the one which has suggested the qualifying adjective *pentata* of the species.

(4) On the abstract hypothesis that the total number of radial canals of the 996 medusae could be equally divided between the specimens (as one would do in partitioning money), we should obtain the "arithmetic mean" of 5·004, which also has, as do all the preceding notions, a probabilistic interpretation.

Which of these four descriptive criteria and of the corresponding parameters — range of variation, most probable value, mode, arithmetic mean — best expresses the "law" of the *Pseudoclytia*? Jeffreys, on the basis of his example, would straightway have replied "five"; but there are several other answers, equally valid. The choice is a pragmatic question to be decided case by case, and the decision has the same subjective basis as with all measurements (section 4346).

We may go further. The proposed solutions are valid for the sample examined, but for another sample the frequencies, probabilities and means would doubtless come out differently; and it is to be remembered, unless contra-indicated, that only the mode[1] would be the same.

---

[1] Anthropology provides a proof in the following percentage distribution of two large samples of human skeletons, distinguished by the number of their ribs:

	According to Melnikoff (per cent)	According to Graffi Benassi (per cent)
11 pairs of ribs	2·6	0·7
12 pairs of ribs	96·8	92·2
13 pairs of ribs	0·6	7·1

Clearly the arithmetic means vary with the percentages, while the mode remains unchanged; accordingly we say that man has 12 pairs of ribs.

While the task of analysing systematically the numerical determinations of the parameters f, g and others can be left to physicists,
this investigation faces the inquirer with quite complicated problems of
probability. However, there will be domains of variation, probabilities,
and arithmetic means varying from group to group; and if no modal
values can be identified (an impossible thing with measurements) certainly one will observe a clustering of the frequencies around the
arithmetic mean.

5621 Laws expressed by mathematical functions

Jeffreys habitually uses the term "adjusted" to indicate calculated
parameters. This term, corresponding to the Italian adjective *stimato*
or *interpolato*[1], indicates a method of analysis which has more in common with the manner of determining and expressing scientific laws
than with the theory of "interpolation". According to the dictionary,
to interpolate means to insert into a text words or phrases which were
not originally there. But the word is used by mathematicians and
statisticians to indicate the substitution, for a complex of particular
quantitative observations, of a general analytical expression based on
theoretical considerations. The technique is usually complicated but,
in general, an interpolatory formula can be reduced to the form $y =$
$f(x, z, \ldots \,|\, k, h, \ldots)$, that is, to a functional expression with variables
x, z, $\ldots$ and parameters k, h, $\ldots$, suitable for characterizing a set of
experimental values of the empirical function Y corresponding to y.
See, for example, section 543, which suggests the expression of daily
variations of marriages by a Fourier polynomial. According to whether
one judges the differences between the observed values of Y and the
corresponding calculated ones of y to be great or small, accidental or
systematic, one can say that the preselected polynomial is suitable, or

---

Also the characteristics most commonly held to be fixed, such as fingers
and teeth, can vary; we even know of human monsters (although they do
not survive) with from 0 to 4 eyes, and higher animals with supernumerary
paws and other organs. But in general the mode does not change. (From
M. Boldrini, *Zibaldone*, Cisalpino, Varese, 1947, p. 49.)

[1] Neither the term "adjusted" nor its Italian translation is included in M. G.
Kendall and W. R. Buckland, *A Dictionary of Statistical Terms*, quoted
above. In the *Multilingual Demographic Dictionary*, prepared by the Inter-
national Union for the Scientific Study of Population, English Section,
United Nations, New York, 1958, p. 8, it occurs only in the usual demo-
graphic meaning of "adjusted rate". In the Italian edition of B. Colombo,
Dizionario demografico multilingue, published by the same Union, Giuffrè,
Milan, 1959, it is translated as "quozienti standardizzati o normalizzati".
The translations given in the text probably convey Jeffreys' meaning.

not, to express the law of daily variation of marriages within the range of time and place considered.

5622   Laws formulated by means of models

Scientific laws, however — as well as the forecasts based upon them — can be expressed not only by interpolatory formulae, but also in "models", that is, by equations or systems of equations (kinds of differential equations) which interpret theoretical considerations derived from experience. The parameters of the model express the interdependence between the variables, and they remain constant during the forecasting period.

Whereas forecasting by means of interpolatory functions is based on certain smoothing coefficients which are invariable with respect to time, the user of models starts from the constancy of the parameters[1]. The formulae concerning the force of gravity and the period of a pendulum are models, in that they provide the basis of forecasts within the tolerated margin of error. In economics, the use of models has gained an increasing prestige in modern times, due to their application in the theoretical study of phenomena, in those branches of economics dealing with incentives and with the control of trade fluctuations, and in many other ways. Model-building has in fact spread to numerous branches of research, both scientific and technical, and is closely studied by workers in many fields[2].

5623   Laws formulated by generalized descriptions

A different formulation of a scientific law consists of a generalized verbal description. This is a relatively primitive method, on which the achievements of such subjects as geography, geology, linguistics, ethnology, etc. have long been established. This method is becoming obsolete, in accordance with an irreversible process generally known as

---

[1] Istituto Statistico delle Communità Europee, "Metodi di previsione dello sviluppo economico a lungo termine", in *Informazioni Statistiche*, No. 6, 1960, p. 575.

[2] Cf. *I modelli nella tecnica*, Accademia Nazionale dei Lincei, Roma, 1956. On demographic and econometric models, see M. Boldrini, *Demografia*, Giuffrè, Milan, 1956, pp. 366, 427 and elsewhere. On analogical and technical models it is useful to read the article by L. Lauriola, "Il concetto di analogia nella scienza e nella tecnica", in *La Scuola in Azione*, 1961–2, No. 14. For economics see the little volume by Adalberto Predetti, *Il Modello di R. F. Harrod*, Isco, Rome, 1961, which refers to an author frequently quoted in these pages.

"from quality to quantity", which, in its turn, expresses a methodological law (section 121). But many branches of natural science, even the most progressive, still abound in qualitative laws. Just think of some well-known instances: "the sun rises in the east and sets in the west"; "litmus paper turns red in an acid solution"; "butterflies keep their wings vertical when they alight, while moths keep theirs flat"; "a body at rest is in equilibrium when the perpendicular passing through its centre of gravity falls within the perimeter of its base"; "widows are more numerous than widowers, because women marry younger than men and live longer".

Reflecting on this and other similar formulae, one understands that, with some of them, it would be difficult to abandon the qualitative mode of expression, while in other cases — as for example the last two — not only can the transition to quantity be imagined, but it is frequently carried out.

Quantitative laws appear to be more peremptory and objective, and they hide the subjective content under cover of their approximative nature[1]. However, just when the opportunity and the method are found for quantifying them, variations emerge, and a model must be chosen within acceptable probabilistic limits (section 4346 and elsewhere).

5624   General considerations on laws

Let us turn now to quantitative laws. The correspondence of a function $f(x, z, \ldots | k, h, \ldots)$, whether it be a model or an interpolatory function, to a set of observations Y is a simple historical statement. A scientific law is something more, and if we wish to proceed schematically, we must recognize that a law should conform to one or other of the following conditions:

(1) It must be possible to bring several samples of a phenomenon, chosen at random, into the mathematical scheme of the single sample on the basis of which the formula was deduced. We have seen in the footnote to section 562 that a second group of human skeletons reveals a modal number of ribs of 12 pairs, identical with the first number; and it was stated in the same footnote that while the mode of the number

---

[1] But even in the qualitative field, formulae are revised from time to time, and thus their subjective aspect re-emerges. For example, for the above-noted difference between butterflies and moths, modern entomologists have substituted another, based on the similarity or otherwise of the nervous systems of the anterior and posterior wings.

of ribs remains unchanged in the two samples (and almost certainly in further samples too) the remaining parameters deducible from the data vary. This is why it is more usual to say that the number of ribs in man is 12 pairs than to calculate the arithmetic mean M, which will probably vary by little or much in successive test groups. The number 12 is, in this case, the parameter of a law, as is the number 6 in the case of the stamens of Cruciferae and in particular of the mustard flower, and as 5 is the number of radial canals in the mantle of certain medusae. However, the parameter is always the result of a decision taken by the observer and is not a natural fact.

(2) When the phenomenon under examination is characterized by a certain formula for a given set of observations, in order that one can speak of a law it is necessary not only that this formula be applicable to other sets (by which condition we go back to the preceding case), but that its use can be extended beyond the observed limits. This condition differs from the foregoing in that it involves the idea of a phenomenon extending in time or space beyond the immediate domain of observation and continuing to behave in the same way. If the supposed Fourier polynomial was shown to be a valid description of the daily variations in marriages both elsewhere and in the future, one would say that this polynomial expressed the general law of the phenomenon. In this research on future confirmation, one must naturally lay down the conditions of validity with great care.

Given a function $f(t)$, found valid in the time-interval t to $t + \Delta t$, it may be that such a function also furnishes valid results because they are equal to or near to reality outside that interval. This use of a functional expression for limits of time beyond those for which it has been derived is called "extrapolation". A formula which is found to be valid for extrapolation evidently expresses a law, although it describes, and is based on, quantitative observations within a limited range. A negative example, well known in economic statistics, occurred between the two World Wars when, following certain studies done at Harvard University, the mistaken idea spread of indices having been discovered forewarning of trade fluctuations. But confidence in this notion vanished as soon as extrapolations from it were shown to be false.

At times an interpolatory function turns out to be valid only in general terms in regard to a set of data different (in space or in time) from that determining it; at other times its validity extends also to numerical values of the parameters. Naturally, in the second case the law is more precise than in the first.

The validity of what has been said about interpolatory functions can easily be extended to models. It is almost superfluous to add that extrapolation and extension can never replace universality, even in the case where all possible occurrences have been exhausted.

(3) One circumstance which even better qualifies some scheme to be a law is when it consists of a mathematical expression which can be used to predict phenomena without depending on their parameters. The famous Einstein models of relativity merit particular mention here; they allowed of forecasting the gravitational bending of stellar light-rays in the proximity of a large mass. The phenomenon was later observed in 1919 by Eddington, during a total eclipse of the sun. Conversely, we may recall Michelson's work on interference, from which it was sought to derive experimental confirmation of the presumed drag of the ether; the non-success of the celebrated experiment of Michelson and Morley supplied an important premiss for the theory of relativity (section 111)[1]. In demography, it is stated that the well-known Verhulst model, built up in expounding the doctrine of Malthus, not only explains the development of populations but can produce useful results in the study of unrelated phenomena such as individual growth, elastic and electromagnetic hysteresis, the development of auto-catalytic chemical reactions, and many other processes[2].

5625   Margin of error in laws

Laws formulated mathematically, therefore, either describe the classified experience or serve to forecast its development in other fields of space or time; or, again, they illustrate the behaviour of quite different phenomena, perhaps not directly observable (section 5511), which are deduced from them or to which they can formally be traced back. The description and forecast are never rigorous but are approximate, so that the worker is always concerned with judging whether the divergences between the observed values of Y and those of y calculated on the basis of the functional formula adopted are such as to lead to the acceptance or rejection as law of the conclusions derived from the formula. Jeffreys puts this point very well[3]: "What we ordinarily do is to adopt some law that fits the observations within a range of what we call 'error'. If we find at some time that the observations cannot be fitted by a law of this form within the range of error that we believe possible, we modify the form of the law. The modified law is then taken as standard until further modification is needed."

---

[1] W. Heisenberg, *Physique et Philosophie*, French edn, Albin Michel, Paris, 1961, p. 119.

[2] See the reference by M. Boldrini in *Demografia*, p. 424.

[3] *Op. cit.*, p. 36.

With these words everything has been said: the approximative meaning of each and every law, even in the physical domain wherein regularities are commonly held to be especially exact; its transitoriness; the subjective inclination of the scientist to modify it every time this appears inevitable. One understands from these words that when Jeffreys, astronomer rather than mathematician and logician, gets down to facts, he is ready to conform to them and to renounce the abstract premises attributed to him above.

5626   Errors and tests of significance

The concept of "random" error is one of the most fertile and complex in Statistics. We alluded to it in connection with the Gauss–Quetelet distribution (sections 381, 4443, 4511, 452 and elsewhere), but modern Statistics has added enormously to the theory, and there now exist whole treatises on the theory of measurement, on the classical "method of least squares", and on the so-called "tests of significance", that is, on the criteria for evaluating the divergences allowable as errors in the sense of the Gauss–Quetelet theorem[1]. The evaluations are always made in terms of probability, and statisticians in particular follow one of two main procedures: either the principle of fiducial intervals or confidence limits (Fisher), or the method of testing hypotheses due to E. S. Pearson and J. Neyman. In both cases the observer's judgement intervenes, and he will avoid any temptation to arbitrary action by imposing an *a priori* limit of probability below which he will reject as invalid the hypothesis on which his calculations are based. Naturally, as it is a question of probability and not of certainty, he always accepts a residual margin of error.

563   Economic justification of the simplest laws

Until now we have not fathomed the depths of this subject, except for what was said in section 56. A phenomenon can be traced back to many mathematical forms or laws; a theory can have many formulations. What will be the criterion of choice between the possible alternatives? The answer is that, as between two or more possible formulations of a classified regularity all equally expressive, and in the absence of valid directives suggested by the data, preference will be given to the formulation which offers the greatest economy of procedure with a

---

[1] A. Naddeo, "La teoria dei tests statistici", in *Quaderni della Scuola Enrico Mattei di Studi Superiori sugli Idrocarburi*, Milan, 1962.

minimum of mental effort and a gain of evidence. Engineers sometimes laugh at themselves when they interpolate a set of points on a diagram with a piece of string — though wrongly, because if by such a simple device one obtains a convenient and plausible expression of the phenomenon better than by a mathematical explanation, then it is technically and logically justified.

This view is the only one which accords with the strict position originating from Locke and adopted in these pages, namely of recognizing the complete inability of science to explain the working of Nature. Mach had clearly reached it (section 311, footnote), the speculations of Poincaré and the neopositivists lead to it, and the most thoughtful epistemologists of today accept it (section 336).

The preference given to simple descriptive summaries will never become a hard and fast rule, since these are conventional in character and recommended by economy of expression. Indeed, in many cases less concise formulations turn out to be more suitable to express the course of a phenomenon and to extrapolate it outside the region of observation. The accurate evaluation of deviations — in particular a decision as to whether they are accidental or systematic — can assist interpretation; and the advantages of a less elementary form of expression accumulate with repetition, allowing the number of parameters to be increased by successive approximation without requiring calculation *ab initio*. Consequently, the advice to choose the simplest solution will not be taken literally, because it does not stem from a general principle but is only a matter of convenience. On the other hand, one should resist the urge to discard it for extrinsic reasons such as mathematical elegance, originality, or the achievement of non-essential side-results, because to do so would be to transgress against the precept of truth, which means — always and only — order, consistency, and practical and cognitive utility.

564   The positivistic and the modern view of laws

Since, in a broad sense, scientific laws are often expressed by interpolatory functions or by means of models — that is, they translate into mathematical terms and generalize the values from a sample of classified observations — one may read with appreciation and respect the following advice written at the end of the nineteenth century by the great astronomer Schiaparelli[1].

---

[1] G. V. Schiaparelli, "Sul modo di ricavare la vera espressione delle leggi della natura dalle curve empiriche", in *Le opere di G. V. Schiaparelli*, vol. VIII, Hoepli, Milan, 1937.

"This [what he had previously said] demonstrates how false is the idea that the progress of meteorology and of some other sciences consists principally in representing phenomena by analytical formulae. Such representation does not advance us one step towards the knowledge of the true law of such phenomena. On the contrary, I think that no-one will charge me with paradox when I assert that whatever formula you wish will never replace without disadvantage the table of numerical values from which it was deduced; this is because in the formula an arbitrary element is introduced, and the greater regularity which its values give by comparison with the observed values is not infrequently the effect of a mis-statement. In fact, it is impossible for the formula to increase by a hair's-breadth the certainty of the observed data and to produce something not contained in them; while it is clear that the fixing of the analytical formula, by no other criterion than the convenience of the calculator, cannot be other than detrimental, introducing into the expression of the phenomenon conditions which are generally not satisfied. . . . Let us therefore use formulae as instruments of research, but do not let us forget that the reduction of a table of observed values to an analytical expression is of no use in itself to our progress in the investigation of the laws which that table represents, except in those very few cases in which such laws have a simple mathematical expression which we can easily predict."

These impressive words bring into sharp relief the divergence between the present-day view and that of positivistic thought. The conclusion conforms occasionally to the current one, but by how much have our thoughts changed since then! Aware as we are today that it is not the task of science to know the "true law" of phenomena, but only that it stops at the *quia*, there is nothing to do but note the usefulness of analytical descriptions which further the aims of science itself. It is true that with analytic expressions of whatever nature — from the arithmetic mean of a series of observations to Newton's law — information is lost with respect to the original data, though lucidity is gained. Furthermore, an analytical table has a historical character, while a mathematical formula, which may extend its validity beyond immediate observation, attributes to some phenomenon a universality which, despite the caution with which we say such things today, is nonetheless always the aim of scientific research.

It should be added that tables of "original data" are not objective reality (as was held in the nineteenth century) but classified truth. Therefore, if the reasoning which refers back to them were to be developed in reverse, we should arrive at the rejection *en bloc* of science itself and of all it can tell us here and now, in favour of history. The negation of science in the name of scientific truth would thus be the nemesis of the proud philosophy of positivism.

5641   Historical character of laws

Speaking of hypothesis, it was said above (section 542) that scientific explanation often consists in tracing back an event which we call "unknown" into the ambit of one or more others already "known".[1] A law shows itself more credible and hence "valid" the more vast is the complex into which it is inserted, and with which it unifies the event or events which form its object. The resulting intellectual gratification, as has been remarked by Pasteur, both delights and astonishes.

However, for the detached observer, it is also important to note that certain unifications widely accepted today, such as those in section 5421, are the modern outcome of earlier, very contrasting conceptions. Hence the observer, reviewing successive scientific epochs, realizes the transitoriness of doctrines which cease to be true as they give way in their turn to others[2].

One could, if needed, adduce further confirmation of the idea that so-called natural regularities and the future development of events do not rest on any objective principle but on induction, as it has been set out in these pages. To the facile objection that nobody has yet seen exceptions to gravity, such as a solid object rising instead of falling, one could reply that a log of wood, thrust down into water, comes up again and floats; but hydrostatic principles and Newtonian theory assure us that the floating is nevertheless a falling (of water in place of the log); without these principles the phenomenon would have a very different appearance.

5642   Heisenberg on the unifying function of laws

The conceptual, and hence non-objective, value of the unification of phenomena which are formally different is made evident in these words of Heisenberg[3]:

"The two laws of the conservation of mass and the conservation of energy lose their separate validity and combine into one law which can be called the conservation of energy or of mass. Fifty years ago,

---

[1] Explanation through unification is different from that by analogy, which can also be fallacious. Cf. A. Maros dell'Oro, *op. cit.*, p. 81.

[2] These contrasting movements can be considered, in an Aristotelian sense, as gravitating towards their respective spheres.

[3] W. Heisenberg, *Physique et Philosophie*, p. 129.

when relativity theory was formulated, this hypothesis of the equivalence of mass and energy seemed a complete revolution in physics, a
revolution then based only on a very few experimental tests. Today,
many experiments show that elementary particles can be created,
starting with kinetic energy, and can disappear as radiations; the transmutation of energy into mass and vice versa no longer appears ab-
normal. The enormous liberation of energy during an atomic explosion
is a new proof, even more spectacular than the exactness of Einstein's
equation."

These words teach us two things: that Einstein's theory has unified
two previously separate concepts, that is, it has made a single law of
two different laws, and this in accordance with the process examined
above; and that it has revolutionized an earlier system of truth, just
as Galileo and Newton had done with the phenomena of mechanics.

5643   Wittgenstein's view of explanation as conceptualization

The conceptual unification to which scientific laws lead, in particular
the one stated above concerning concepts already proposed by Newton,
recalls a well-known passage, unusually detailed and definite, in
Wittgenstein's *Tractatus* (para. 6.341). "Newtonian mechanics reduce
the description of the universe to a unitary form. Let us imagine a
white surface on which there are irregular black spots. Let us say:
whatever figure may result from it, I can always approximate as much
as I wish to its description by covering the surface with a square mesh
sufficiently fine, and saying whether each little square is black or white.
I should thus have reduced the description of the surface to a unitary
form, a form which is arbitrary, since I could with equal success have
applied a net with a triangular or hexagonal mesh.

"It may be that the description with the triangular mesh would be
simpler; this means that with a coarser triangular mesh we could
describe the surface more exactly than we could with a finer square
mesh (or vice versa), and so on. The different nets correspond to
different descriptions of the universe. Mechanics determines one form
of description thus: all propositions in the description of the universe
must be obtained in a given manner from a given number of other
propositions, the axioms of mechanics. It thus procures the bricks for
the construction of the edifice of science and says: whatever edifice you
wish to build, you must always build it with these bricks and only
these. (As with the system of numbers one should be able to write any
number, so with the system of mechanics any proposition of physics.)"

This idea is developed in Wittgenstein's next paragraph and concluded in the following one (6.343). "Mechanics is an attempt to con-

struct according to a single plan all the *true* propositions which we need for describing the universe."

It is enough to notice the italicized word *true*, to understand how much Wittgenstein shunned the old idea of objectivity. For him, scientific laws are form: "Laws such as the principle of sufficient reason, etc., are concerned with the network, not with what the network describes" (6.35). Other similar statements form the subject of the axioms which follow. (For the non-reducibility of all physical laws to mechanics, cf. sections 111 and 3326.)

In conclusion, to "explain" means to unify conceptually; law is not an essential property of Nature but the inductive formulation of its actual behaviour.

565   Summary notes

The theme we have been developing can be summarized under the following heads:

(1) The observational phase, from which is derived the high-level hypothesis, presupposes the whole theory of induction. The formation of statistical "cases" — apparently so simple when it is a matter of counting the eyes and fingers of man — becomes of primary importance when (to keep to our example) it is a question of counting the eyes of an insect, the flagella of a protozoa, the cells of an organ, etc. Naturally, there is greater skill and greater risk of error in the quantitative determination of phenomena which are measurable but not enumerable, especially where the observations necessitate the use of delicate instruments and involve considerable choice of interpretation.

(2) The formulation of a law always has a deductive character, and this is reaffirmed by experiments suggested by it.

(3) Neither the highest-level hypothesis nor the intermediate ones depend upon general principles of Nature of the type stigmatized in section 561, but only upon the immediate premises or upon those of previous theories from which the highest-level hypothesis itself can eventually be derived. This circumstance allows absolute liberty to the researcher, who will judge, case by case, the value and generality of the law derived from its own or other premises, and will abandon it when it is no longer adequate to the system in which it is applied or to the theoretical or practical needs from which it arose.

(4) It follows from the discussions in section 3221 *et seq.* that no scientific law can be absolutely confirmed through experiment. However often we synthesize water (as was said earlier) we shall never arrive experimentally at the universal proof that two atoms of hydrogen and

one of oxygen combine to produce a molecule of water. What, then, is scientific truth? Let us return to the earlier examples. What do the following laws mean: "the mustard flower has six stamens", "the Pseudoclytia has five radial canals in its mantle", "$g = 9·80$ m/sec^2"; "$f = k(mm'/r^2)$"? Formerly the answer would have been that they are objective determinations having their ultimate basis in the perpetuity of causes and in the eternal laws of Nature. It has been repeatedly said that the modern answer is quite different. Science postulates no universal principle such as causes and laws of Nature, and its axiomatic premises derive their validity only from the useful deductions made from them. The regularity of Nature is created by the mind which, from the observed facts, produces by induction instances, common nouns, and statistical data, aligning them in a timeless succession and assuming them to be permanent until, as such, they become of service to man. A law, then, is true when groups of events inductively homogenized, co-ordinated, and consistent with other laws and systems, can be embraced within it without contradiction. Its validity is essentially transitory and historically conditioned, because it changes with the development of knowledge. The nature of the generalizations of uniformity now reveals, in a certain sense, an economic element which makes progress alongside the purely intellectual one. In today's cultural climate, in which science and technology are increasingly linked, that side of scientific knowledge advances which is most sensitive to the urgency of satisfying both the highest intellectual needs and those concerned with practical living.

Neohumanism, of which we spoke elsewhere (section 261), finds confirmation once more in the modern attitude of the scientist towards Nature.

566 Democritean and Machian orientation in science

An opportunity to recapitulate the concepts we have developed comes through a cultured and penetrating writer[1], who recognizes in contemporary science two antagonistic attitudes which he calls "Democritean" (alluding to Lorenz, Rutherford and Planck) and "Machian" (Mach, Poincaré and Carnap). The first group favours a modern gnoseology because it relies only on principles to which an objective knowability can be attributed (e.g. water being constituted of hydrogen and oxygen, colour being the expression of electromagnetic waves with a length of between 0·4 and 0·7 micron).

---

[1] A. Maros dell'Oro, *op. cit.*, p. 73.

As for the second group, says dell'Oro, "confusing the means (models, measuring instruments, systems of co-ordinates, etc.) with the end (to know Nature, where experience does not penetrate), they have claimed theory to be completely conventional." They are pure experimentalists, and according to them "it is unscientific to try to enlarge our knowledge with suppositions in regions where direct experience cannot reach."

Dell'Oro, who, as noticed in section 5511, distinguishes between theory and laws of Nature, avoids either of these extremes. The point of view chosen in these pages is nearer to Mach than to Democritean atomism, but we have also refrained from assuming a concept of science "emptied" (as dell'Oro says) "of all intrinsic value, in that theory could give nothing new". Such a position would be appalling, because the justification for all research resides in every new discovery that it achieves.

One asks, then, is there something beyond immediate experience? Dell'Oro criticizes physicists such as Heisenberg, Born, Dirac and Jordan who uphold this point of view. But it is clear that they consider as valid the theories set up by themselves. Beyond immediate experience there is the complex rationalization that man makes of it, and this is what constitutes science. It consists neither of a catalogue of direct experiences nor of arriving at immutable laws of Nature.

567   Rejection of dualism

But, leaving the abstract, let us first of all make clear that — at least in these pages — we have never maintained that science consists of its instruments, but that it is, on the contrary, man's interface with the external world of which he is a part. Such contact is neither intuitive nor immediate nor crude; it does not consist in looking and describing, but reveals itself methodically, beginning with information on entities, passing on to events through classification, and building up to a structural arrangement by mental processes involving mathematics and ordinary language. Science, we have argued, is a mental construct of a slice of the world, but not invention; it is the arranging of the singular in the plural; it is the interpretation, subjective but coherent, of this plural; it is pictured as having discipline, linked on the one hand to all knowledge — even to what is remote from ordinary experience — on the other to human needs both intellectual and material; finally, it is a linguistic expression based on the common noun.

One might maintain that as mathematical (and also qualitative) expression is the final product of every scientific inquiry, whoever re-

J

ceives the message it conveys might see in the formulae and words no more than an instrument: hence the justification of dell'Oro's criticism of those whom he calls the "modern Machians". But that would be greatly to lower the intellectual value of science. Science, while manifesting itself and making itself known by its language, remains a genuine human contact with the world, and hence consistent and cognitive. In affirming this, one excludes the idea of detached objective knowledge, and hence the idea of reality gathered, so to speak, passively. Every scientific affirmation is a working out of events according to a plan which Nature helps to suggest but whose truth man determines according to his classificatory criteria, the establishing of analogies, their consistent application in other contexts, and finally by mathematical treatment which attributes to science a formal rigour in no way different, except in content, from that of the enunciations and theorems of pure mathematics.

Even in science one could speak of tautologies, taking care to avoid misunderstandings about the factual content always present in applied mathematics. But the justification of the proposed analogy is precisely this: that pure and applied mathematics have the same formal characteristics and always draw their conclusions from antecedent ones, whether these are abstract or not.

568   A warning from Jeans

We have said and repeated above that deductions drawn from hypotheses are compared with observational results, but that these latter cannot give confirmation of universality. "Nature", wrote Sir James Jeans some years ago, "can reply to our question by showing us a phenomenon which is in contradiction with our hypothesis, or a phenomenon which is not in contradiction with it. It can never offer to our observation a phenomenon which proves it. One phenomenon is enough to destroy a hypothesis; but a million million are not enough to prove it." Very true, except perhaps for the excessive emphasis on "destroy". It would have sufficed to say that a phenomenon that does not fit into the framework requires a hypothesis to be modified.

Let us, then, take the warning of the celebrated physicist from his fund of truth, recognizing that the experimental control of a hypothesis is always indispensable as providing uncontradicted and hence provisional confirmation of the hypothesis and of its deductive derivations.

571   Statistics in scientific structure

It is on this terrain that science and Statistics are interlinked and mutually integrated. Perhaps the reader of the last few pages may have reason to feel somewhat perplexed. What is the position of Statistics? From the concluding part of Chapter 4, at any rate, the distinction between logic and Statistics seems clear; to the first belongs the task of connecting the propositions in a "subject," to the second the functional linking-up of the "subjects" into scientific structures (section 481).

A structure is the application of a set of rules by means of which one comes to synthesize a complex of natural facts. We have examined the threefold process involved, considering the phases one by one; and at the end we asked ourselves whether, given their cyclic nature, it was really a question of three phases or rather of three aspects of the same procedure, all of which combine to characterize the structure. If this is so, the constructive unification of scientific procedure in its formal universality is neither physics, biology, meteorology, economics, linguistics, nor art criticism; but neither is it logic, because it transcends the individual subject and becomes Statistics in its highest meaning.

When one emphasizes the complexity of the concept of structure — to be understood as involving in practice an interrelation of the syntactic operation of Statistics and the semantic knowledge of individual natural sciences — the inadequacy of Carnap's concept of Science, already referred to in section 17, becomes apparent. For this most able logician, the physics of a given epoch is a system of propositions formulated in natural language; and the propositions having a universal form are called physical laws. Carnap gives an analogous definition for biology and its laws. For this to be true, it would be necessary to leave out of consideration all the complex of operations which are disentangled in the threefold scientific process.

The structure of science consists in this mutual operational dependence. Hence the physics (or biology) of a given epoch must be seen as a discourse on a group of phenomena observed, worked out and interpreted in accordance with the semantic canons of contemporary physics and the permanent syntactic rules of Statistics. Only when it passes from the general to the particular and is inserted into the context of applied disciplines can Statistics take on that customary look conferred upon it by "cases", "data" and "tables", and, through mathematical language, become an efficient working instrument.

How does this welding come about between the diversity of the particular sciences and the unifying effect of Statistics, understood in this sense? This book has not been written to contrast the answers found in the different technico-mathematical systems. But when one considers, even very generally, the natural development of knowledge, it is easy to see how a working agreement is reached between science and Statistics, whether this is determined by the natural scientist or by others. From this point of view, Statistics becomes of the greatest possible importance, and its field of application extends to all branches of science. Some years ago I wrote a short sketch of the history of Statistics which makes this evident, and a later comprehensive work by dell'Oro[1] brings out the point no less strongly.

572 Hypothesis, deduction, induction, and Statistics

In Statistics, a hypothesis is information; it must, therefore, originate from outside, from the scientist who formulates and extracts it from his own studies. As has already been noted, the hypothesis often reveals the originality, the mental concentration and at times the genius of its originator (section 541). Perhaps for this reason stories and legends abound concerning the origins of hypotheses which were the prelude to great discoveries. Pythagoras is said to have discovered harmonies while listening to the alternating sounds of smiths' hammers on an anvil. It seems that Archimedes had an intuitive feeling about the principles of hydrostatics while in the bath. The smoking stove of Descartes, the swinging lamp of Galileo, and Newton's apple are also universally quoted incidents. One could wish that Lagrange thought of the calculus of variations while listening to an organist playing variations in the cathedral of Turin, and that the movement of a lift helped Einstein to consider relative motion. Again, chance suggestions led Röntgen to the discovery of X-rays, and Fleming to that of penicillin[2].

Hypothesis, however, always originates from the observation of numerous cases (since a single fact decides nothing), and numerous classified cases are already a statistical fact[3]. Here, then, is a starting-point where the scientist's invention and classificatory enumeration stimulate one another. From the hypothesis is developed the chain of deductions; in the empirical sciences, however, these are closely inter-linked with technical ideas and processes. Only a chemist can know what reactions to look for when combining certain elements; and only an economist is capable of indicating the effect of a monetary or banking measure. Many other similar situations could be suggested. But if the experimental plane concerns the specialist, the statistician will indicate the general strategy, the amount of research, and will guard against errors of fact and interpretation.

---

[1] A. Maros dell'Oro, *Storia della Statistica* (1967). See also the second chapter of my *Statistica, teoria e metodi* (1942).

[2] M. Boldrini, *Zibaldone*, p. 38.

[3] *Ibid.*, p. 173.

Finally, the collection and interpreting of observable elements will make up the third phase, inductive documentation. Based on experiments carried out by the scientist, statistical techniques will then suggest to him or to his collaborators how to deal with and generalize the observations, and how to compare these with the hypothetical model and adapt this to tally progressively with the results, so that from the initial inadequate or incorrect form there emerges a definite conclusion, consistent with the premises and capable of interpretation in the widest possible context. In a word, it indicates to him how to reduce this conclusion to express the law of the phenomenon studied.

573   The technique of Statistics

As has been said above (sections 511 and 571), we had no intention of going into detail concerning the technique of Statistics, which nowadays has become a colossal subject, not to be contained in one or even many volumes. There are numerous works to which reference may be made (such as those of Fisher, Kendall and Stuart, Wilks, Vianelli, and Boldrini), but in themselves they do not show how to initiate the simplest inquiry, if the person using them lacks the creative gift. Strictly speaking, when confronted with a new problem, it would in fact, be contradictory to look for a means of resolving it among predetermined procedures which were necessarily suggested by quite other needs; while it is clear that if the problem is an old one, it can only represent a misfortune for whoever unknowingly takes it on. Admittedly the technical methods of the manuals are set out in general terms; they correspond to certain types of problem and are in a certain sense adaptable; but precisely for this reason they require in whoever uses them great experience, inventiveness, know-how in adaptation, and caution in avoiding errors.

An analogy can be drawn from therapeutics. Every patient is a case, but many similar cases suggest the course of the illness, and experience has taught how to treat it. The doctor has at his disposal numerous medicaments already prepared, which of themselves would be unsuitable if he were not capable of administering them, watching their effects, combining them with others, and sometimes himself writing a prescription. Likewise, in scientific inquiry, pre-existing formulae and procedures are not always enough; on the contrary, to attack the most unusual problems it is necessary to invent new procedures, and this is the privilege of creative minds. Experts know how much modern Statistics — to be precise, that of the school of Fisher — owes to the needs of agricultural experimentation, and how much that of the earlier school of Karl Pearson to the study of the problems of evolution. But Karl Pearson and Fisher are eminent examples of original minds, from whom every statistician should draw inspiration if he wishes to reap from his own work — as theoretician, scientist, or collaborator in experimentation — those genuine rewards which will increase, even if modestly, man's knowledge and command over Nature.

5731   Two fundamental lines of development: Bayesian and non-Bayesian techniques

Naturally, statistical technique has kept abreast of the times, and nowadays two lines of development are open for the posing and solution of its problems; these are connected with questions previously treated.

De Finetti[1] points out that with the first line of development, called "Bayesian", there is assumed *a priori* an initial distribution of probability (that is, the hypothesis) in order to reach, by the use of Bayes' theorem, the distribution of final probability (that is, the "law"). Adherents of this procedure are statisticians and probabilists who accept the "subjective" theory of probability (de Finetti, Savage, etc., sections 42, 4343 *et seq.*) or who assume Bayes' theorem in an axiomatic form (Jeffreys, section 4333). According to these authors, starting from the "principle of coherence", the information obtainable initially about a phenomenon is sufficient to formulate an *a priori* valuation of its probability, and hence to proceed to the solution of the relevant practical problems.

The second, "non-Bayesian" line is favoured by those statisticians — and they are the great majority — who give an "objective" definition of probability, whether based on propositional relations or on likelihood (section 42).

Yates[2] has written that Fisher "during the whole of his life con-

---

[1] B. de Finetti and L. J. Savage, "Sul modo di scegliere le probabilità iniziali", in *Biblioteca del Metron*, series C, "Note e Commenti", vol. 1, Rome, Istituto di Statistica della Facoltà di Scienze statistiche, demografiche e attuariali dell'Università di Roma, 1962, p. 82.

[2] F. Yates, "Sir Ronald Aylmer Fisher (1890–1962)" in *Revue de l'Institut International de Statistique*, vol. 30, No. 2, 1962, p. 280. Very important for the questions concerned are the pertinent criticisms of M. G. Kendall, "Ronald Aylmer Fisher, 1890–1962", in *Biometrika*, **50**, 1963 [reprinted in *Studies in the History of Statistics and Probability*, ed. E. S. Pearson and M. G. Kendall, Griffin, London, 1970].

tinued to develop his ideas on the fundamental problems of inductive inference and estimation. He soon recognized the fallacy of the Bayesian approach," and in 1930 Fisher introduced an original concept which others, better known than he, followed up during his long scientific career. Not only Fisher but nearly all the leading modern mathematical statisticians, such as Neyman, E. S. Pearson, Bartlett and many others consider that only on the basis of elements derivable from objective distributions can one arrive at scientifically valid statistical estimates.

For reasons mentioned above, in this study we do not have to list, much less to formulate, the relevant theories. On the contrary, to remain on epistemological terrain, it must be said that the dichotomy proposed by de Finetti is excessively drastic; the reasons for this judgement have already been implied in section 4345. In fact, at the root of every scientific problem there is always a subjective element, but research sets out with the instruments and methods best adapted to making the reported results objective and unequivocal. In the calculus of probability, even the subjectivists need to set up the "principle of coherence" as an axiom in order to formulate logically and linguistically the theorems derived from it. This is equivalent to imposing constraints on opinions and to opening the way to a rational — that is, in a certain sense, an objective — development of the theory of probability. Nevertheless, it remains equally true that some statisticians accept Bayes' theorem as an instrument of research while others do not; thus the antagonism is shifted from the rarefied atmosphere of principles to that which surrounds a pragmatic choice — which I, also, decide in a non-Bayesian sense, though with a mind always ready for any innovation that may be required by scientific consistency.

5732 Anscombe on relative domains of application

Rather than reject either concept of probability — the experimental, which he calls "chance", or the subjective, for which he reserves the classical name of "probability" — Anscombe puts forward two possible fields of use. "When the choice turns on the procedure in a decision to be used impartially in an automatic way (as, for example, a project of public-opinion sampling by a Government department), the analysis made by an orthodox statistician may be good and pertinent: the Bayesian statistician will accept such an analysis and can add very little to it. On the other hand, when the problem concerns a single intelligent decision (such as, should this new idea be actively developed or set aside?), it is not clear that the orthodox statistician's analysis can have much deciding force; often only the Bayesian can enlighten and assist the workings of common sense."[1] Two alternatives, therefore, though expressed in a guarded manner; but later the author seems to decide on the validity of the second, illustrating what he considers the advantages of the Bayesian method by an experimental market-research sample inquiry in which an industrialist proposes to establish the suitability or not of producing and marketing a new product.

5733   Faleschini on objective probability in Statistics and subjective probability in operational analysis

Although Faleschini's exposition is persuasive, it must be objected that by far the most general and pertinent case which faces the statistician is not the second but the first of the alternatives proposed. The distinction between the two cases (as has been mentioned in section 4347) is well brought out. According to Faleschini, as between the two probabilities, objective and subjective, only the first involves statistical technique, while the second appertains to a branch of applied mathematics which sometimes becomes assimilated and confused with the former through inaccurate terminology. The aggregates constituting the world, as we have already said, are either historical, H, or of a classificatory nature and can obviously be described as statistical, S. The probabilities of the S naturally have an objective character, like the laws into which they enter or with which they are combined. Therefore, according to Faleschini, in the theory and technique of statistics there is no room for subjective probability.

But, he adds, besides the cognitive activity, which tends to the construction of S aggregates and hence of scientific laws, there is also a technical or operational activity which, starting from the knowledge of scientific facts s, singles out a subset of s in a certain historical aggregate H and uses it in a given factual situation to modify H into H'.

We recall from an earlier section that the probability $P(H'/H)$ is essentially an indeterminate relation of the type $0/0$ to which it is sometimes possible and useful to attribute a value on the basis of known information. Probabilities of this type occur in operational practice and

---

[1] F. J. Anscombe, "Bayesian statistics", *The American Statistician*, February 1961. Against the ideas maintained here, see I. D. J. Bross, "Statistical dogma, a challenge", *The American Statistician*, June 1961. In various senses, for the most part opposed to the neo-Bayesian, these problems have been discussed in the same review by A. Birnbaum (February 1962), J. W. Pratt (April 1962), I. D. J. Bross (February 1963) and H. O. Hartley (February 1963).

their use is peculiar to so-called "decision theory". Clearly, decisions are by nature individual and thus are unique, but they are based on scientific information and hence on repetitive data. Whether to span a river H with a steel, concrete, or masonry bridge, taking account of the width, the solidity of the banks, the load to be transported, the cost of materials, the time necessary, and all the scientific data relating to materials and supplied by engineering knowledge — is a decision which the engineer must take for that particular case. In the same way, whether to cure Mr H of a gastric ulcer surgically or clinically is also a decision, taking note of the gravity of the case, the general conditions, the results of examination, the knowledge of anatomy, physiology, pathology, etc., as well as the confidence placed in the doctor and surgeon[1]. Finally, included in the same category is the market inquiry hypothesized by Anscombe, with the help of which an industrial director should be well informed before deciding whether to produce and market a new product.

These and all other imaginable decisions depend on the subjective value attributed to the corresponding probabilities, and naturally to the reliability of the scientific information to hand. It emerges from the above examples that when, in such circumstances, we speak of "statistical decisions" we are using the adjective inaccurately, because the decisions are concerned with specifically historical situations and not with solving general problems.

To conclude: the positions of de Finetti and of Faleschini are basically different, but the latter also speaks of subjective probability, although he limits the validity of the concept to certain historical situations which are extraneous to Statistics — in which, however, de Finetti definitely includes it. Furthermore, Faleschini tends to agree with Anscombe's position, but confirms the preference to be given to the orthodox statistical solution, because this, and only this (although surrounded by the precautions developed in section 4346) constitutes the instrument with which the scientist systematically faces his prob-

---

[1] There exists a vast international literature on operational research. Besides the important volume by H. Chernoff and L. E. Moses, *Elementary Decision Theory*, John Wiley and Sons, New York, 1959, note also the article by M. Boldrini, "La classificazione delle scienze e le analisi operative" in *La Scuola in Azione*, No. 22, 1963–4, which is developed along the lines summarized in the text.

lems, starting from binding premises, and tries to reach general and communicable results[1].

58   Classification of the sciences: from Bacon to Comte

In this book we have repeatedly referred to the separation into classes of scientific subjects, according to their method of research and their content, and in section 51 (as frequently elsewhere) we have used the term "natural sciences" to indicate those which classify, co-ordinate, elaborate, and explain linguistically specific groupings of events. Underlying these references and nomenclature there is a question of considerable epistemological importance — that is, the "classification of the sciences".

Discussion of this subject started in the seventeenth century, that is, at the dawn of modern science, and has continued, with affirmation and contradiction, almost until the present time. Francis Bacon, the first to theorize on induction, distinguished the sciences according to "the three parts of the human intellect", that is, memory, imagination and reason; and on the basis of these faculties he recognized three fundamental classes of knowledge—history, poetry and science. He

---

[1] The uniqueness of operational analyses, supported by the intuition of whoever intends to use them, contrasts with the systematization of scientific inquiries in the not infrequent cases in which the former are carried out so speedily as to show themselves only in the end results, without allowing a rational reconstruction of the process. There is no doubt that in every difficult technical decision there lies an evaluation of the probability of success, together with the cultural background of the person concerned. But it often happens that a general or an airman cannot give a convincing account of the rational and improvised manœuvre which has saved him. Similar circumstances sometimes occur with discoveries in those scientific fields in which so-called intuition (which is a quick rational analysis of the factors involved) can play a large part.

How R. A. Fisher discovered the sampling rules governing symmetrical functions having cumulants as means was a mystery even to himself. "Fisher", relates M. G. Kendall, "was never able to explain it to me in later years. . . . He was on a train journey when he thought of the combinatorial rules which express the cumulants. . . . When Fisher came to write the paper, however, he had forgotten how he arrived at the basic results . . . and his 1929 paper concludes with an obscure section which is no proof at all. . . . Authors occasionally have a mystical feeling, on looking back, that some of their work was written through them, not by them." It is a question of essentially unique results, realized perhaps with the rapidity of an urgent operational analysis, not dissimilar to the way in which a surgeon decides on immediate treatment almost at the same time as he works out the diagnosis.

surveyed the first two categories rapidly, but thoroughly developed the third, in which he distinguished a *philosophia prima*, a *theologia* and a *philosophia naturalis*. To the last, the celebrated treatise of 1605 on the progress of culture, re-written in Latin in 1623, is almost entirely devoted[1].

From the beginnings of science to its powerful position in modern times the path is via the celebrated *Cours de philosophie positive* by A. Comte. This opens with the "discovery" of a fundamental law, that is, of the three "theoretical" states of every branch of our knowledge, which are the "theological or imaginary", the "metaphysical or abstract" and the "scientific or positive". Comte considered the scientific phase as the culmination of "the progressive march of the human spirit". He recognized also that "in phenomena of every order two kinds of natural sciences must be distinguished — the one group abstract, general, and having as object the discovery of the laws which regulate the different classes of phenomena, all possible examples being considered; the other group concrete, particular, descriptive, sometimes designated by the name of natural sciences, and consisting in the application of these laws to the actual history of various living entities".

From such a basis, Comte visualized in his penetrating way the six famous classes of the sciences emerging — astronomy with mathematics, physics, chemistry, biology and sociology, arranged in this order — that is, of increasing complexity and with a progressive use of the inductive method[2].

581   Croce's objection to the classification of the sciences

Incidentally, Croce points out that the two classificatory systems just mentioned are not isolated lists, but essential categories in the philosophies of Bacon and Comte[3]. However, after admitting this, he denies to his contemporary philosophers the right to do likewise, and hence rejects the legitimacy of various attempts which were being made

---

[1] F. Bacon, *The Advancement of Learning and New Atlantis*, Oxford University Press, London, reprinted 1960, pp. 81, 82, 96, 100 onwards. The Latin edition of the work of 1605, appearing in 1623 with the title *De Dignitate et Augmentis Scientiarum*, gave the author the opportunity to enlarge it and to make further divisions between the sciences and introduce considerable changes.

[2] A. Comte, *Cours de philosophie positive*, 2nd edition by E. Littré; J. B. Ballière et fils, Paris, 6 vol., 1864. The fundamentals can be read in the first volume, while succeeding volumes are dedicated to the six branches of science. The last quotation in the text is from page 56 of volume 1.

[3] B. Croce, *Logica come scienza del concetto puro*, Laterza, Bari, 1909, p. 263.

at the beginning of the present century to build up reasoned classifications — that is, philosophically admissible ones — of the branches of
scientific knowledge. And here he deploys the vehement irony which
he so often used successfully to disarm his adversaries. When faced with
certain useless endeavours, he says, one should refrain from "discussing
them philosophically so as not to waste time in misunderstandings, or
at least to realize their unimportance. . . . The pen falls from the hand
and the head is bowed in meditation at the sight of someone who
prefers not to put aside this dangerous game and occupy himself with
something else. . . . Of the poet Aleardo Aleardi it was said that,
addressing the Muse at every instant in his poems to ask her for something, he treated her as though she were his waitress. The professor
ends by treating Science as his janitor, or at least as his worthy consort,
with whom he amicably agrees about the dishes to be chosen for the
day's dinner and about the other business of running a family."

Croce must be excused because he was sometimes embroiled in the
rough-and-ready metaphysics of late Positivism, but one cannot agree
with his denial to others of a right which he claimed for himself when
affirming the reality of the categories of the mind, that is to say, of the
particular scientific domains in which he himself had faith.

582   Reply to Croce

Croce should have asked himself why so many able men, at the end
of the nineteenth century, took so much trouble over such sterile
research; for since the classifying of the sciences was less pleasurable
than whiling away an evening playing patience, it must indeed have
answered some real need. Even in Italy in 1904, A. Ravà, not in fact
a positivist philosopher but rather a neo-Kantian, put forward a classification of the sciences; and the need seems to have revived in the climate
of the present-day renewal of scientific philosophy.

583   The natural sciences in the classification of the sciences

The ancient distinction between formal (or abstract) and applied
sciences (section 145) has often been presented in these pages, and
among the applied subjects the historical ones, namely those directed
to the study of individual entities, have been clearly separated from
those in which natural facts are reduced to events and classes.

Since the formal or abstract disciplines are usually known as sciences
of the "universal", the classifying ones (not the historical) might be
given the name of sciences of the "general", if they did not have to be
subdivided into two groups, to the second of which this designation

applies less than to the first. Such are the "observational" or "natural" sciences or, as dell'Oro calls them, "real" — and those which are "operational" or "arts". The name of the first of these groups, repeatedly used in this book, derives from their use in the construction of theories, projects, and models for ordering natural facts and, with the help of mathematics, eventually reaching regularity and relations (scientific laws) between the phenomena concerned. On the other hand, the name of the others, of mediaeval origin, refers to the fact that, using ideas either from history or the natural sciences, men derived the rules applicable to individual practical ends. Botany is a natural science, and the statement that the mustard flower has six stamens is one of its "laws"; and engineering is an "art" which is based on mathematics, physics and technology (section 5733).

We mentioned above that Comte, the founder of Positivism, used the expression "natural sciences" in a wider meaning; but it is also true that when he included mathematics among them, he placed it alongside astronomy, in which it was almost engulfed.

However that may be, here we wish to underline the distinction between formal and natural sciences. It is generally agreed today that logic and mathematics are included among the first, but one ought at least to add Law, for the sake of its theory, whose nearness to logic is made clear from its origin and also from the tendency already noted (section 173) to use in it the language of logistics[1]. Perhaps one should not stop here. As for the natural sciences, the definition just given evidently covers all those subjects which by long tradition have been designated in that way, as well as many others which have been forming during modern times, together with older branches of knowledge

---

[1] The origin of formal Law in logic is shown in the first treatise on civil law in which Q. Mucius Scaevola, taking his inspiration from Greek models, laid the foundations of the dogmatism and co-ordination of the Institutes. "What the Greeks had done in philosophy", wrote C. Ferrini (*Il Digesto*, Hoepli, Milan, 1893), "and particularly in rhetoric, he wished by a happy thought to carry over into law." And the famous specialist in Roman law added: "For the rest, his definitions are built up according to the formal precepts of logic; he first defines the main genus, and then (usually in a negative form) goes on to narrow the field until he arrives at the required species." The method has not changed in the course of centuries. In the eighteenth century, Beccaria asserted the logical consequentiality which must characterize law, indicating "geometrical precision" as its method. And in our day, C. Magni writes thus (*Avviamento allo studio analitico del diritto ecclesiastico*, Giuffrè, Milan, 1956, especially p. 74): ". . . certain legal relations, hitherto considered as created by legal research, become particular cases of relations derived by formal logic".

which were known in the past under a different guise. Ancient and undisputed natural sciences then, are, physics, chemistry, zoology, botany, mineralogy, meteorology, and others; and new sciences are demography, genetics, anthropology, biometrics, cybernetics, and others. But what can we say of systems which, as problems and research methods alter, have approached ever nearer to the natural sciences? Psychology with its many branches, linguistics, and philology have gradually ceased to be the study of individual cases and, once started on the road of classification, have likewise succeeded, in varying degrees, in mastering the quantitative and the mathematical instrument.

There are, it is true, many natural disciplines in which quantification has scarcely yet appeared, but in these the systematic use of classified entities foreshadows a more or less imminent passage from quality to quantity. Such are physical geography, geology, and the wide range of social sciences from ethnology to sociology properly so called.

584   Statistics and the natural sciences

Now a common characteristic of the natural sciences, understood in this broad sense, is the considerable use of the statistical method in research procedure, a use which is formally established in the more progressive branches. The use is, however, only potential in those sciences where the researcher has to deal with unmeasured quantities such as much and little, great and small, rare and frequent, dear and cheap, hot and cold, and others of the same kind (section 181). However, in section 4624 we also recognized the existence of non-numerical statistics in which such terms are substitutes for measured quantities.

Here, then, is a precise definition of the sciences called natural, which follows on logically from the complex of ideas developed in this book: the natural sciences are those which use Statistics explicitly, or at least potentially, for research and the formulation of relations between the phenomena classified[1].

585   Composite disciplines

Separate examination is needed for certain branches of knowledge which, viewed from one angle, show a deductive structure, but from another, present a different aspect. It concerns, in fact, the sciences that combine under a single name two main bodies of doctrine which are juxtaposed: the one definitely formal, the other naturalistic. Thus, "pure" economics — that is, deductive and mathematical — is in-

---

[1] For a fully documented statement see A. Maros dell'Oro, *op. cit.*

cluded in the science of economics, as also are the various branches of applied economics, with their specific names of political economy, economic statistics, econometrics, and others.

Analogously, physics, says Max Planck, is theoretical when it produces an image of the world representing, up to a point, a mental construct; on the other hand, it is experimental when it describes the external world which is revealed by the senses or by measuring instruments which function as perfected sense-organs.

The position is similar with applied mechanics and so-called theoretical mechanics. Linguistics, in its turn, remains a historical subject while it studies the origins, forms, and diffusion of languages; but it becomes a natural science — according to Schleicher's concept (1863), modernized, freed from its Darwinian trappings, and still accepted today — as soon as it sets out to compare and collect them into classes and families, and to search for their laws of formation and change.

Of art criticism we have already spoken (section 355) and some may think that there we went too far.

Finally to deal with the second of the classes of "general" sciences, mentioned in section 583, come the operational disciplines (corresponding, more or less, to the *Artes* of the Schoolmen), whose content is derived from the formal sciences and from various technologies. First in the line is medicine, whose principles are founded on biology, psychology, physics, chemistry, and other subjects; then engineering, already mentioned, finds its scientific premisses in physics, chemistry, mineralogy, and hydraulics, as well as in mathematics; accountancy, which depends on arithmetic and applied economics; and lastly the military art and agricultural experimentation, which are also built up on knowledge from multiple sources.

Clearly, then, Statistics has a contribution to make to the structure of composite disciplines, although in a limited degree.

586   Autonomy of Statistics in relation to the natural sciences

The characterization of the natural sciences on the basis of their statistical structure is of great importance, apart from what we have just said, because it is the key to the further understanding of the autonomous position of Statistics in many natural domains and in the working procedures connected with them.

What, then, is the position of Statistics in the classification of the sciences? It is easy first to say clearly what Statistics is not. First and foremost, it is neither a particular science with its content of facts, as the earlier theoreticians had believed; nor is it a system composed of ideas, as is true of some of the examples mentioned in the preceding section. Earlier writers argued at length on these points. German authors towards the end of the nineteenth century thought of Statistics as a social science, or rather, in John's phrase, as "definitely the realistic guide of the social sciences".[1] More alive to international trends, and agreeing with Luigi Bodio — who, some years later, was to be the most active among the promoters of the International Institute of Statistics — Angelo Messedaglia, eminent among the exponents of the old school, was to slough off the qualitative and descriptive trappings and to envisage Statistics as "the science of social facts and of their laws, through homogeneous groups of quantitative elements".[2] No less peremptorily, Gabaglio wrote later of the science that "studies the politico-social order of facts through quantitative observation".[3]

At the beginning of the present century, A. Kaufmann, Professor at St Petersburg (Leningrad), also revealed the progress made in the subject by beginning a treatise with these words: "Statistics is above all, as Bruno Hildebrandt has already defined it, a 'political and social science of measurement'; according to a more recent definition by Lexis, it is the exact numerical inquiry into the manifestations of human society. Intuitively we can define Statistics in a lucid way as 'a method of research which tends to clarify the systematic collection of quantitative observations by the comparison of numerous typical groups' (Conrad)." Therefore it is a social science, although, adds the author, "not *only* social, because its field is much wider".[4]

But when it was recognized that neither the social sciences nor any other subject was identifiable with Statistics, its methodological nature became apparent, as the pioneer Benini showed when he defined it as a "method of observation and induction" (1906). The concept was, however, still limited, because Statistics was recognized as a method only in those disciplines whose phenomena "present themselves as a plurality or mass of cases susceptible of variation without a strictly

---

[1] V. John, *Geschichte der Statistik*, Part 1, F. Enke, Stuttgart, 1884, p. 14.

[2] A. Messedaglia, "Della statistica e dei suoi metodi", in *Archivio di Statistica*, 1, 1876; "Della scienza statistica della popolazione", *ibid.*, 2, 1877.

[3] A. Gabaglio, *Teoria generale della statistica*, 2 vol., Hoepli, Milan, 1888, vol. 2, pp. 8, 15. Besides this work, which at the ttime of first publication was an exemplary treatise on the history and technique of Statistics, one may also mention the modest little volume of F. del Prato, *Sinossi di storia e teoria generale della statistica*, Fiaccadori, Parma, 1880.

[4] A. Kaufmann, *Theorie und Methoden der Statistik*, J. C. B. Mohr, Tübingen, 1913, pp. 1, 5.

assignable rule". This was correct: but it was said with the intention of excluding the physical sciences, since it was commonly agreed that they recognized only deterministic phenomena, not variable ones. That was an error which the physicists themselves explicitly corrected. A pertinent statement by Max Planck may be quoted: "In every physical law, even in gravitation and electrical attraction, there is a nucleus of a statistical nature, and we are always concerned with laws of probability based on average values of numerous observations which are of the same type and in single instances are only approximately valid."[1] These points should by now be fully appreciated by the reader.

587   Statistics is a formal science

Statistics, therefore, with the characteristics we have been delineating, is also one of the formal sciences (section 481), and must be added to the incomplete list of these given above (section 145 and 583). A near relative of logic, Statistics is not identified with it, since, while it uses the methods, of logic, it organizes them to form the triple procedure of hypothesis–deduction–induction. Furthermore, Statistics is linked with mathematics because it uses its numbers, algorithms and methodology, but, again, is not identified with it inasmuch as the entities it deals in are concrete not abstract, and stand for classified, measured phenomena and not merely ideas. It is further differentiated from both logic and mathematics because statistical relations never belong to the domain of certainty — even when for algorithmic reasons they might seem to do so — but to that of probability. And because of this, some people tend to identify it absolutely with the calculus of probability — but wrongly so, because that would represent an obvious limitation.

It might seem strange that the comparatively new name of Statistics was not bracketed long ago with older subjects as a formal science. But this oversight is not unjustified, because, even in recent times, statisticians have continued to qualify their science as a "branch of logic", and still more recently (as we have seen) have confused it with the calculus of probability. Naturally, its nearness to Law, to the extent to which this is related to logic, is more incidental than substantial. In the practical field, the points of resemblance and of difference are of relatively little use, but in a theoretical treatment such as the present they could not be neglected. This is all it seems possible to say at present to put into perspective the compendious but inadequate defi-

---

[1] M. Planck, "Die Kausalität im Naturgeschehen", in *Scientia*, **43**, 1933.

nition with which the preceding chapter concluded, so as to detach the Theory of Statistics completely from traditional disciplines and raise it to its proper status in present-day epistemology.

59   Taking leave

Science in the twentieth century no longer concentrates its attention on the unattainable essence of things but creates its own structures, having their own basis of truth and internal consistency; and yet the scientist, if he once abandons the cold impartiality of research, again discovers in himself those lively and constant springs of action, so much more complex and compelling than strict reason, which are called interests, affections, temptations, conflicts, victories, and impulses for good. The analysis of these emotions one by one can constitute the subject of scientific theories, but in their entirety, at once harmonious and inharmonious, they only make sense and come alive in the secret places of the heart. A truly scientific book ought not to concern itself with this aspect: but it is something that no man, not even the scientist, can ignore.

The last sentences of the famous *Tractatus* of Wittgenstein, who has given the impetus to much thinking in the last half-century, bring support to this view when, losing their axiomatic detachment, they break out into pronouncements and questions full of anguish and mystery. "The solution to the enigma of life in space and time is to be found outside space and time" (end of 6.4312). "A superior being is completely indifferent as to how the world may be. God does not reveal himself *in* the world" (6.432). "There is in truth the inexpressible. It *shows* itself, and it is this which is mystical" (6.522).

The final aphorism, although vaguely suggested earlier (6.5), appears as an act of will, a will perhaps tired and uncertain, because it imposes silence about things of which one cannot speak: "Wovon man nicht sprechen kann, darüber muss man schweigen" (7). The logician in him, it is true, could not speak: but he had rejected the claim to eternal truths of his early Catholic education, though still, perhaps, inwardly influenced by them.

I too (and forgive me for comparing myself with that great man) have striven, during a long working life, to confine my own researches within strict limits of method freely accepted. Of late years I have been concerned with the gravest thoughts, but now that I have arrived at these last lines, I take up again my own interests as a human being and look with faith beyond the dark wall which will close the last day, convinced that the ineffable mystery waiting there is called God.

Principal References

BENINI, R., *Principii di Statistica metodologica*, UTET, Turin, 1906.

BOLDRINI, M., *Statistica, teoria e metodi*, Giuffrè, Milan, 1942 (4th edn, 1962).

BOREL, E., *Valeur pratique et philosophie des probabilités:* vol. 4 of *Traité du calcul des probabilités et de ses applications*. Gauthier-Villars, Paris, 1939.

BRAITHWAITE, R. B., *Scientific Explanation*. Cambridge University Press, 1955.

CARNAP, R., *Logical Foundations of Probability*. University of Chicago Press, 1950 (2nd edn, 1962).

DE FINETTI, B., *La probabilità et la statistica nei rapporti con l'induzione, secondo i diversi punti di vista*. Cremonese, Rome, 1960.

GEYMONAT, L., *Filosofia e filosofia della scienza*. Feltrinelli, Milan, 1960.

HARROD, Sir Roy, *Foundations of Inductive Logic*. Macmillan, London, 1956.

JEFFREYS, Sir Harold. *Scientific Inference*. Cambridge University Press, 1957.

KENDALL, M. G. and STUART, A., *The Advanced Theory of Statistics*, vol. 1, 1958 (3rd edn, 1969), Griffin, London.

KEYNES, J. M., *A Treatise on Probability*. Macmillan, London, 1921.

VON MISES, R., *Wahrscheinlichkeitslehre, Statistik und Wahrheit*, Vienna, 1928· English edn, *Probability, Statistics and Truth*, Allen & Unwin, London, 1955.

MORTARA, G., *Lezioni di statistica metodologica*. Tip. Leonardo da Vinci, Città di Castello, 1922.

NAGEL, E., *The Structure of Science*. Harcourt, Brace & World Inc., New York, 1961.

PEARSON, K., *The Grammar of Science*. Dent, London, 1937 (1st edn, 1892).

RUSSELL, B. (Lord), *Human Knowledge: its Scope and Limits*. Allen & Unwin, London, 1948, Parts I–IV.

SAVAGE, L. G. et al., *The Foundations of Statistical Inference*. Methuen, London, 1962.

Index